THE CANAL PIONEERS

THE
CANAL
PIONEERS

BRINDLEY'S SCHOOL OF ENGINEERS

CHRISTOPHER LEWIS

The
History
Press

To Sophie, Emma and George

Canals lay at the centre of economic life, creating a transport revolution and assuaging a fuel famine, or energy crisis, even more severe than that of the twentieth century. They influenced the growth of population, reduced widespread unemployment, provided extensive passenger services, aided the agricultural revolution, and served as the main arteries of heavy industry for about a century. No wonder the men who built them, realising something of their potential, became obsessed with the canal idea.

Hugh Malet, 1977

Frontispiece: The former Armitage Tunnel on the Trent and Mersey Canal, now opened out.

First published 2011

The History Press
The Mill, Brimscombe Port
Stroud, Gloucestershire, GL5 2QG
www.thehistorypress.co.uk

British Library Cataloguing in Publication Data.
A catalogue record for this book is available from the British Library.

ISBN 978 0 7524 6166 3

Typesetting and origination by The History Press
Printed in Great Britain
Manufacturing managed by Jellyfish Print Solutions Ltd

Contents

Acknowledgements

Thanks are due to the generous help and advice provided by Mike Chrimes at the Institution of Civil Engineers, as well as to Carol Morgan, in the archive library at the Institution, for her help. Thanks are also due to the Warwickshire and Staffordshire Record Offices, the Potteries Museum and Art Gallery at Hanley, the William Salt Library in Stafford, Keele University Library, the National Waterways Museum, the Public Record Office, the British Library and the libraries of Birmingham and Sevenoaks.

Of the many people who have given advice and help on the writing of the book, particular thanks go to Peter Cross Rudkin and Chris Rule for correcting misapprehensions on my part and, on more occasions than I like to acknowledge, my Brindley-like syntax and spelling. I am also greatly in debt to John Norris and Sue Hayton who have helped most generously with researches on Thomas Dadford and Samuel Simcock respectively.

Thanks go to members of my family: George Lewis for the design of the cover, assistance with digital photography and the loan of a camera, Emma Young for the drawing of the maps, Kam Young for computer work and Sophie Stevens for her researches on the Chesterfield Canal and help with the gazetteer. Thanks also to William Harris for help with the gazetteer and instructive trips to the Montgomeryshire and the Wilts and Berks canals. Thanks are also due for work, undertaken long ago, by the late John Lewis, who interpreted the labyrinthine will of Hugh Henshall and gave advice on genealogy. Also to the late Charles Hadfield, who was more than generous in the lending of his personal notes on the engineers; that was a different age.

Additional thanks go to Hugh Taylor, Barry Jackson and Gary Johnson for their whimsical and philosophical company on various canal trips, and to Danny Hayton for his picture of Brindley used on the cover. A final thank you goes to Malcolm James and Sally Kelly who kindly and patiently copied photographs and dealt with the mysterious workings of the computer world.

Christopher Lewis
Sevenoaks, 2011

Introduction

School: 'System of doctrine as delivered by particular teachers'
– Dr Johnson, 1755

The 'Eighteenth Century was seized by a rage for order', according to Henry Hitchings. This manifested itself in everything from charting the globe; standardising weights, measures and time; erecting signposts on highways; using account books; creating dictionaries; and, of course, building canals to link strategic ports, rivers and the areas of burgeoning industry and agriculture. This 'rage' enabled Britain to develop into the first industrial country of the world, and the 'canals became the midwives of this new age'.

Two engineers stand out in the middle of this optimistic, industrious and supremely energetic century. The first was John Smeaton. Formally educated, Smeaton was a brilliant mathematician, instrument maker and builder of lighthouses, harbours and river navigations whose eponymous Smeatonian Society eventually led to the creation of the first Society for Civil Engineers in Britain. Indeed it was Smeaton who coined the name civil engineer, to differentiate between the previous work of military engineers and the work of the new engineers developing the infrastructure of Britain.

The second was James Brindley. James Brindley was the pioneer in the vanguard of the practical age of canal construction in Britain. Unlike Smeaton, Brindley was untutored, some claim illiterate, and began his work as an apprentice millwright. However, it was Brindley who developed and laid down the principles of constructing our early canals.

The laying out and building of what was, in effect, the beginning of a national system of canals was too great for one man, however. Here, Brindley's vision and organisational ability was revealed. He created what Cyril Boucher called a 'School of Engineers' to reflect his designs, draw his maps, survey territory under his direction and build the canals he was commissioned to construct. At the time Brindley was working, there were no engineering schools in Britain that were similar to the French schools (the first of which opened in 1747 to teach surveying, levelling and engineering as an academic discipline). Brindley instead learnt in the field, drawing towards him men who worked and learnt with him too. Some of the engineers and surveyors in his 'school' came from land agents who had been involved in managing great estates; Hugh Henshall, later to become Brindley's brother-in-law, fell into this category.

The men who became Brindley's pupils were in some cases, and in some areas, technically more competent. Yet whilst they worked with him to realise his designs and translate them into a practical reality, they did not have his *idée fixe* as to the building of canals and they did not possess, initially, the knowledge of hydraulics and canal construction particular to Brindley. Some might have drawn better maps, built longer tunnels or created successful businesses connected with canals, but they lacked the overall idea of the revolution in water transport, and they lacked Brindley's ability to persuade businesses of 'the canal idea' at the outset of the Canal Age.

This book is an account of the lives and works of these little-known engineers who began their careers working with James Brindley. Between them they constructed the original canal system in England, Wales and Scotland. It would be wrong to term them pupils only in the literal sense, although they have been referred to as such and the phrase 'School of Engineers' implies a tutor-and-pupil relationship. This was not the entire picture. Certainly at the outset, Brindley laid down working practices in canal construction that he had developed on his projects, and others followed them. Later it became a more co-operative venture; a school in the doctrinal sense. Brindley co-operated with, as well as instructed, the engineers described in this book, and he trained them to his own high standards in canal construction. Many of the engineers went on to extend Brindley's original system, known as the Great Cross, as the country came to realise the considerable benefits that canals conferred in the transportation of heavy goods and materials.

Hugh Henshall, Samuel Simcock, Robert Whitworth, Josiah Clowes and Thomas Dadford were colleagues, friends and relatives who all belonged to the Brindley school of engineering. To write an account of their lives and achievements has been difficult. The eighteenth century left little evidence of their existence. It has been, in one sense, an archaeological exercise; scouring canal company minute books, where they exist, and examining occasional letters exchanged between the members of the school, fellow engineers and company managers. This, plus evidence from parish records of births, marriages and burials, has slowly built up the picture. Wills and gravestones also yielded evidence of these men, their families and lives. All, barring Brindley, left wills; some had made considerable monies from their labours, while others died in penury.

The absence of information about their characters is frustrating. We have no portraits of these engineers by brush or pen as we do of Brindley. We do not know their stature, their colouring or their temperament. In some we may infer reliability; in some, business acumen; in others, great interest in mathematical and philosophical pursuits; and in some, sharp practice, frustration and tragedy. These engineers all worked together in a rapid voyage of discovery in the second half of the eighteenth century. Some, such as Hugh Henshall and Robert Whitworth, clearly had the benefit of a good education. The others, as L.T.C. Rolt has said, were men who 'began work as tradesmen and finished it as engineers'. In this patrician phrase, Rolt might have been thinking of Samuel Simcock and Thomas Dadford, or indeed even of James Brindley himself.

To navigate this book, follow the briefest of chapters on canal construction to the individual chapters on the engineers themselves. These chapters focus on the work they did with Brindley and the work they continued to undertake following Brindley's premature death in 1772. Several of them were related to James Brindley and also to one another; this is illustrated in the family trees. The inclusion of the five engineers that worked with Brindley, and then went on to work on the expanding canal system elsewhere, is not of course the complete tally of Brindley's working partners. Many others came within his orbit. Of these, John Varley, Samuel Weston and James Brindley

Jnr – who was James Brindley's nephew – are mentioned in a brief, final chapter. Recent research has uncovered important new facts about Brindley's nephew James and his canal construction in America from 1774 onwards. Meanwhile the major canals which the individual engineers from the Brindley school were chiefly responsible for, are described within the chapters of the book that bear their engineers' names.

Each chapter contains information about the lives of these engineers: where they were born, educated, married and where they lived and died. At the end of the book there is also a gazetteer listing each engineer, their canals and some of their engineering achievements such as flights of locks, tunnels and aqueducts. It lists what other evidence on the ground still exists of these men: for example, buildings where they lived and churches where they were married and, eventually, buried.

The life of James Brindley has been well chronicled in the past four centuries. In the eighteenth century the *Biographia Britannica* published an account of his life based on the reminiscences of Hugh Henshall, his brother-in-law, and Josiah Wedgwood and Thomas Bentley, his friends. This evidence was subsequently used as a basis for Samuel Smiles' intriguing account of Brindley's life in *Lives of the Engineers*, published in the nineteenth century. This account bore the unmistakable message of self-improvement for the reader. In the twentieth century, Cyril Boucher then published an interesting engineering study of Brindley, complementing and clarifying some lesser-known facts about his work and life. This has been added to in our own century by Christine Richardson's detailed biography that attempts to show us James Brindley, the man. She reminds readers less of the excessive parts of Brindley's character which colour some earlier accounts of his life, but highlights his moral strengths as well as his engineering skills. In this she has been assisted by Kathleen Evans' account of Brindley's family background and Quaker forebears.

Although Smiles asserted that Brindley liked to work alone, this is not borne out by the evidence we have of his collaboration with and direction of his school of engineers. He might have liked to analyse matters on his own, but his practical working life was with the support of others, whom he chose carefully and then worked closely with. This book therefore examines those with whom he worked and their importance. Those who share work also share ideas, and there must have been a cross-fertilisation of thought throughout the Brindley school.

Portrait of James Brindley.

Eighteenth-Century Canal Construction

Canal: 'Any tract or course of water made by art'
– Dr Johnson, 1755

The Land Prepared

After a proposal for a canal had been drawn up by a committee or group, following advice from an engineer or surveyor, the committee had to estimate costs, find finance (usually through floating shares in a company) and seek approval from Parliament.

Any application to Parliament for permission to build a canal required a plan and section surveys to be prepared of the proposed route. This is where the services of a good

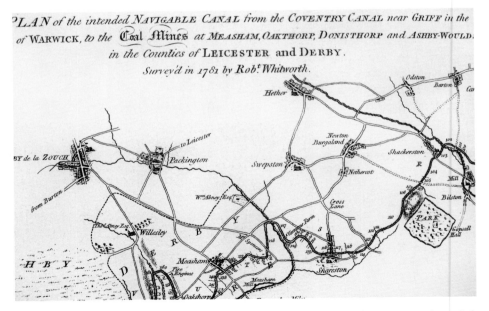

Robert Whitworth's map of an intended canal from the Coventry Canal to coal mines at Measham *et al*.

surveyor and engineer were required to outline salient details and answer any questions posed by Parliament or objectors to the scheme. Sometimes, in the early days of canal building, variations in a line would be made without reference to Parliament.

After approval had been granted an Act of Parliament would be passed and two copies of the plan, certified by the Speaker of the House of Commons, were produced. One was lodged with the clerk of peace for the county where construction was proposed, the other with the clerk of the canal company.

Purchase of Land

At this point, assuming the act was passed, the engineer was informed of the exact bounds within which the law had confined his operations. He then marked out the canal on the ground with the surveyor, if that was a different person, so that the clerk to the canal company could prepare contracts for the purchase of property.

The clerk checked the titles to the lands and arranged conveyance and purchase. Any differences between the landowner and the canal company were adjudicated by commissioners to ensure a fair price, if that was the point at issue, or the sheriff of the county. Land for canals was purchased outright, unlike the system of 'wayleaves' adopted where land was rented from the owner for the construction of early tramways.

Construction

The principal or consulting engineer, or sometimes the clerk of works, then traced accurately the levels of each pound, or reaches of canal between locks, and marked the ground with levelling pegs every two or three chains. Where there was a change of level in the land, the depth which needed to be dug was calculated through triangulation with a staff, or any other marker such as a rock or tree, and a chain. The distance and angle thus measured would determine the depth to be cut. Pegs then marked the top of the cut, and 'slope holes' the intended water level. Spirit levels, telescopes, chains, scales, a plane table and, later, theodolites were used for this work.

James Brindley statue,
Coventry Canal basin.

Where locks were required for the cut to climb or descend a gradient, lock pits were dug and levels continually checked. Developments in the production of instruments in the seventeenth century had a profound effect on the accuracy with which surveys could be accomplished. The English theodolite was introduced around 1720 but was cumbersome, expensive and difficult to carry. Most of Brindley's canals were surveyed before the act was passed, after an 'ochilor servey or recconitoring' by Brindley himself, riding his faithful mare, and by techniques such as triangulation.

Top soil was removed and the banks apportioned; as a general rule, 1ft in depth would give a horizontal base of 1.5ft. The banks of the canal would be 1ft higher than the level of water intended to stand in them. Soil was then removed to form banks.

The resident engineer or clerk of works would let certain lengths of canal for cutting to contractors or hag masters, who would employ a number of navigators under their direction to dig and then seal the canal with puddle, a kind of earth mortar. In the case of large or difficult contracts the principal engineer was called to evaluate work and costs. Such contracts could be let individually to engineers, as with Thomas Dadford on Sow Aqueduct on the Staffordshire and Worcestershire Canal.

The clerk of works was responsible for all matters relating to the supervision of works. Other than responsibilities already mentioned, the clerk kept an account of the hours worked by contractors and, according to Peter Cross-Rudkin, he was responsible for 'measuring and valuing the work of the contractors and keeping all the financial records'. Finally, 'he disbursed all of the cash'.

The canal was then invariably sealed with puddle. Of puddle, Rees says: 'it is a mass of earth reduced to semi fluid state by working and chopping it with a spade, while water just in the proper quantity is applied, until the mass is rendered homogeneous, and so much condensed, that water cannot afterwards pass through it, or but very slowly.' Puddle would not have been left to dry out as it would crack, but would be gradually built up to the required depth, 2–3ft depending on the ground. The canal banks required puddle sealant as well as the bed.

Once the section of canal was completed and puddled, water could be admitted to a depth of 1–2ft and dirt boats could be floated on the thus far completed canal to transport spoil to build-up embankments where required.

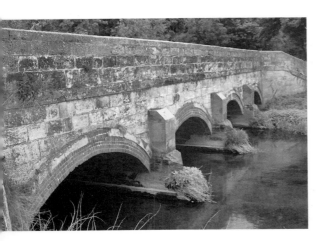

Sow Aqueduct, built by Dadford, Staffordshire and Worcestershire Canal.

Lock at Cropredy, Oxford Canal.

Locks

Locks were used to raise boat traffic up or down from one level to another. Early locks on river navigations were flash locks, where the river was banked up behind planks or staunches at a certain point before some central planks were removed and the boat would cross the difference in level in a flash of water. The boat would have to be hauled by horse, man or winch if proceeding against the flow of the river: clearly, a risky and inefficient procedure.

The first European masonry lock with top and bottom gates, known as a pound lock, was used in 1488 on the River Brenta near Padua. Shortly afterwards two canals in Milan were joined by six locks, similar in design to the pound lock of the present day. James Brindley's first lock was constructed at Compton on the Staffordshire and Worcestershire Canal, but pound locks had already been built on the Sankey Navigation and other river navigations in Britain. For example, locks were built on the Calder and Hebble Navigation, where Brindley had been an engineer in 1765 for a period, after completing the first Bridgewater Canal and before starting work on the Staffordshire and Worcestershire Canal.

Locks were constructed by excavating a lock pit and then erecting a wooden frame. The side walls were puddled before masonry walls were built and gates hung. It was important in setting out the canal to ensure that the fall of each lock was equal. Sometimes, during early canal construction by the Brindley school, this did not always happen. When the falls were uneven then flooding or water shortage would result. The procedure to calculate the fall from each lock necessitated very careful surveying.

Cuttings

The Chinese had built cuttings up to 80ft deep in their very early canals. Important factors in construction included headings below the level of the canal to ensure adequate drainage to combat collapse. The angle of the cutting depended on the strata through which the canal was taken. Drainage headings were also constructed behind the banks created. If the cutting had a vertical depth of more than 60ft, tunnelling became the preferred option. During excavation horse gins could be used as well as barrow runs, anticipating the work on railway cuttings to come. John Carne invented a deep-cutting machine pulled by a horse which was used under Josiah Clowes in his work at Cofton Hackett on the Worcester and Birmingham Canal, and again by Clowes on the Herefordshire and Gloucestershire Canal.

Embankments

These were often constructed from soil removed from cuttings and tunnels. James Brindley used wooden caissons as a basis for building embankments on the Bridgewater Canal. Towpaths were crucial so that horsepower could be utilised in hauling boats. Burnt coals were used to reinforce the towing paths of the Bridgewater Canal and great attention was also paid to controlling moles and rats. If these rodents burrowed into canal banks and workings they could cause serious problems. In 1795 a mole or rat burrowing, followed by a hard frost, occasioned a rupture of the Croydon Canal embankment of 100yds and a boat was deposited on the meadows below. Companies would employ rodent catchers to guard against such catastrophic events.

To minimise the effect of a breach in the embankment, stop gates were also placed at regular intervals so that part of the canal could be sealed off whilst repairs in the embankment were effected. Culverts, to ensure that water did not build-up on either side of an embankment and threaten its stability, were built of wood and brick through the base, thus ensuring suitable drainage so that the pressure of water either side of the embankment did not lead to failure.

Aqueducts

When building aqueducts, secure foundations would be sought and, where necessary, piles sunk to gain a foothold on a sound base. This

Lock near Gargrave, Leeds and Liverpool Canal.

Bollin Embankment,
Bridgewater Canal No.2.

was particularly important where the aqueduct crossed a river to prevent the effects of scouring. The arches of the aqueduct were to be of equal equilibrium to offset the settling of stone or brick and prevent collapse.

In plan, a masonry aqueduct was ideally designed curving inwards, that is, the ends are wider than the middle. The walls should not be upright but battered or diminishing upwards with the outside to give greater strength and stability to the structure. This technique evolved through the errors in aqueduct building which members of Brindley's school sometimes experienced: for example Samuel Weston at Gowy on the Chester Canal. In the case of Kelvin Aqueduct on the Forth and Clyde Canal, additional buttressing was provided to support the sides given the depth and weight of water it carried.

Materials of the best quality were to be selected and the work was always to be completed in the summer season. After completion, the aqueduct was puddle-lined. A fine example was Dove Aqueduct on the Grand Trunk Canal, built by Hugh Henshall.★

All the engineers from Brindley's school of engineering constructed masonry aqueducts. The use of iron came later with Thomas Telford, Benjamin Outram and William Jessop.

Tunnels

Tunnels had been used for mining and drainage purposes before the construction of canals. The first tunnel for navigation was built at Malpas, near Beziers, on the Languedoc Canal in France in 1681. The tunnel, which was unlined and mined from either end, was constructed by Pierre-Paul Riquet and his engineer, François Andréossy. The first canal tunnel in England of significant length was constructed by James Brindley at Harecastle.

★ The Trent and Mersey Canal's original title was The Navigation from the Trent to the Mersey. It became known colloquially as the Grand Trunk Canal. The canal is referred to in the text as the Trent and Mersey Canal or the Grand Trunk Canal.

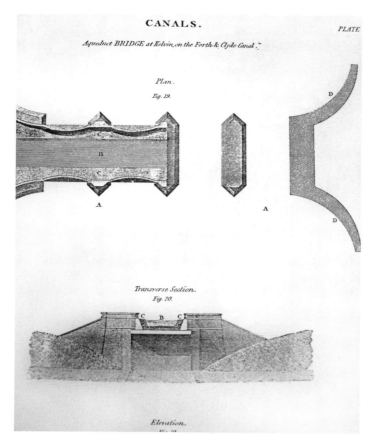

Plan of Kelvin Aqueduct, Forth and Clyde Canal. (By kind permission of ICE)

Kelvin Aqueduct, Forth and Clyde Canal.

Josiah Clowes became the expert within the school of engineers to prosecute major tunnels following Brindley and Henshall's work at Harecastle and Hartshill, or Norwood, on the Chesterfield Canal.

As Rees explains, a line in the exact vertical plane was fixed and staked out over the hill. Secondly, a valley or brook below the level of the tunnel was sought so that drainage

Dove Aqueduct, Trent and Mersey Canal, near Burton-on-Trent.

from an adit could take place. Headings were dug below the planned level of the tunnel so that this drainage could be facilitated.

The line was set out parallel to the tunnel over the top of the rise. Excavation then took place from either end of the tunnel and from vertical shafts leading to headings constructed in the middle of the tunnel. Staffs, sometimes with flags, were erected and then, from a fixed point, the line and depth were calculated in the same manner as when the canal went through undulating ground.

Tunnel pits were dug, the depth being calculated trigonometrically in the manner described previously. String or a fine line was stretched across the top of the tunnel pits and plumb-bob lines suspended from this to check the depth. From the tunnel pits, which were 150–200yds apart in most cases, men and tools could be lowered by windlass or horse gins and spoil removed. Excavation could then take place in line with the longitudinal axis of the tunnel.

Tunnel pits were normally square and lined with wood if no bricks were available. If the tunnel pit was directly over the tunnel arch, damage to the crown of the tunnel could potentially occur. From these pits, headings were dug in either direction until they joined. These headings were then widened to the required dimension of the tunnel, and brick arching was installed where necessary.

Once the direction and depth of the tunnel had been calculated the miners could orientate their excavations underground. The headings or passages were dug from the tunnel pits and supported by timber or brickwork if there was no rock to support the tunnel arching.

Where great quantities of water were encountered, such as at Harecastle Hill, steam engines were installed to pump water from the workings. The steam engine at Harecastle was an early Newcomen engine.

Above: Interior of Sapperton Tunnel, lined.

Left: World's first canal tunnel, Malpas, near Beziers, France.

Below: Spoil heaps, Sapperton Tunnel, Gloucestershire.

Ribs or centrings were constructed by a carpenter, as in Clowes' case, using his 'model or driving frame'. This was thought to be a template to ensure the tunnel dimensions. Soil excavated from the tunnel was to be moved away, possibly to be used elsewhere on the canal, and the tunnel pit was used as an airshaft where necessary. There are fine examples to be seen of the twenty-five spoil pits over Sapperton Tunnel, in Gloucestershire, on the Thames and Severn Canal. After excavation, the walls of the tunnel were lined with brick of varying thicknesses, as required, before water was let in.

1

James Brindley

'A Workman Genius'
– Erasmus Darwin

It is universally acknowledged that James Brindley was a natural engineering genius. Clearly, he was also a man of great imagination with considerable powers of oratory and great charisma. It is not always agreed, however, how much he was personally involved in some of his projects. Sir Joseph Banks, the eminent botanist and later president of the Royal Society, described him 'as a man of no education but of extremely strong natural parts'. It would also be fair to observe, after the passage of 250 years, that he was one of the founding fathers of mechanical and civil engineering; if not its pre-eminent member with regard to hydraulics during the middle of the eighteenth century.

Born in the remote and high hills of Derbyshire, at Tunstead in the Parish of Tideswell in 1716 – a year after the first Jacobite rebellion had been suppressed – James spent his formative years on the roof of England with its river valleys and limestone watercourses. A memorial exists on the site of the cottage where James Brindley was born. The trees growing out of the site were referred to as 'Nature's own Memorial to the Great Engineer', by W.E. Gladstone. James received little formal education other than a possible short period at Overton Bank School in Leek (not proven), and that which his mother provided, plus the work on his father's farm.

Young James became interested in things mechanical at an early age and there were reports from his childhood of his making model mills. In 1726 the family moved to Leek where James was employed on odd-jobs and general farm worker. Aged seventeen, Brindley was apprenticed for seven years to Abraham Bennett, wheelwright and millwright at Sutton near Macclesfield. James' powers of problem-solving and memory are legendary. The young apprentice walked one weekend from Macclesfield to Smedley, near Manchester, to solve the problem of one of Bennett's mills that wasn't working due to the latter's inaccurate design. He observed the mill in Manchester, calculated what needed to be done to the machinery of the mill and walked back to Macclesfield to be at work on Monday morning with the solution: a remarkable feat of memory as well as engineering acumen.

In 1742 Abraham Bennett died and James Brindley, now aged twenty-six, returned to Leek and began his own business as a millwright in Mill Street. From that year until 1765 he lived with his family at a house just outside the town at Lowe Hill. This was the home of his parents and Quaker grandmother, Ellen Bowman. Lowe Hill Farm was also where Brindley's younger brother Joseph later lived with his son, also James Brindley. Brindley was later to work with his namesake nephew on some English canals before James Brindley Jnr went to America to build canals there in 1774.

Brindley's work at this stage comprised repairing agricultural machinery, carts and working mills. For the next eighteen years Brindley was to be employed in a wide range of engineering work in the North Midlands, Shropshire, Lancashire and Yorkshire. He was involved in making and repairing various mills, including ones producing silk, corn, paper and flint. In the case of the latter he was responsible for introducing the first mill, Machin Mill, for grinding flint in water after calcination. In making waterwheels and other engineering components, it is noteworthy that Brindley selected his own timber, felled it and cut it up himself. Gladstone would have approved.

Memorial to James Brindley at his birthplace, Tunstead, Derbyshire.

Much has been made of Brindley's illiteracy. However, in addition to the commentaries already available two points are worthy of consideration. Firstly, his spelling would have been regarded as eccentric as he spelled words phonetically. But it should be remembered that it was not until the middle of the eighteenth century, by which time Brindley was in his mid-thirties, that an attempt to standardise spelling was started by Dr Johnson in his *Dictionary of the English Language*. Johnson also gave guidance as to the correct pronunciation; again not of much help to a man from north Derbyshire with a strong accent. Secondly, Edmund Burke estimated that the reading public of Britain by the 1790s, those who would use or buy a dictionary, was below 100,000. This was not the figure of those who could read and write in order to communicate, but it does place Brindley's writings and calculations – and they were only rough personal notes in his

Brindley's watermill, Leek.

THE BRINDLEY FAMILY

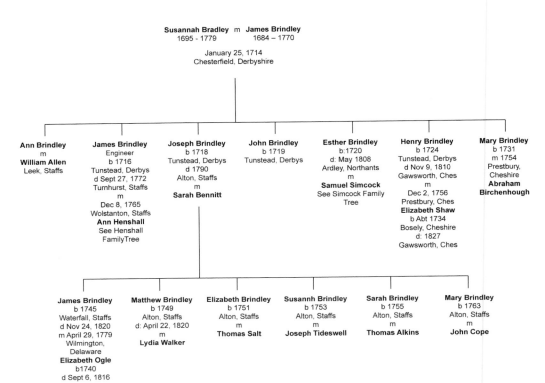

Susannah Bradley   m   James Brindley
1695 - 1779   1684 – 1770

January 25, 1714
Chesterfield, Derbyshire

Ann Brindley
m
William Allen
Leek, Staffs

James Brindley
Engineer
b 1716
Tunstead, Derbys
d Sept 27, 1772
Turnhurst, Staffs
m
Dec 8, 1765
Wolstanton, Staffs
Ann Henshall
See Henshall
FamilyTree

Joseph Brindley
b 1718
Tunstead, Derbys
d 1790
Alton, Staffs
m
Sarah Bennitt

John Brindley
b 1719
Tunstead, Derbys

Esther Brindley
b:1720
d: May 1808
Ardley, Northants
m
Samuel Simcock
See Simcock Family
Tree

Henry Brindley
b 1724
Tunstead, Derbys
d Nov 9, 1810
Gawsworth, Ches
m
Dec 2, 1756
Prestbury, Ches
Elizabeth Shaw
b Abt 1734
Bosely, Cheshire
d: 1827
Gawsworth, Ches

Mary Brindley
b 1731
m 1754
Prestbury,
Cheshire
**Abraham
Birchenhough**

James Brindley
b 1745
Waterfall, Staffs
d Nov 24, 1820
m April 29, 1779
Wilmington,
Delaware
Elizabeth Ogle
b1740
d Sept 6, 1816

Matthew Brindley
b 1749
Alton, Staffs
d: April 22, 1820
m
Lydia Walker

Elizabeth Brindley
b 1751
Alton, Staffs
m
Thomas Salt

Susannh Brindley
b 1753
Alton, Staffs
m
Joseph Tideswell

Sarah Brindley
b 1755
Alton, Staffs
m
Thomas Alkins

Mary Brindley
b 1763
Alton, Staffs
m
John Cope

memorandum, or Day Book, as he called it – in a very different light. Brindley was prudent enough to select Robert Whitworth and Hugh Henshall to help with his map-making and reports to Parliament. Also, there is no doubt that he could convey his knowledge and ideas to everyone; from members of the House of Lords to clerks of works, labourers and masons at the canal side. Paradoxically, James Brindley was a great communicator: witness his effect on Members of Parliament, in conversations with contemporaries such as Josiah Wedgwood and with men of his School of Engineers.

In 1750 Brindley realised that the growing industrialisation of the Potteries (around Stoke-on-Trent) was such that more work would be available further north and so, in 1750, he rented a workshop from Wedgwood in Burslem. He then worked from the two centres, including work repairing machinery for Josiah himself. This act was to change the course of his career. Until this time he had been what might be termed, in today's parlance, a mechanical engineer. His arrival in the Potteries led him further north to begin looking at civil engineering projects under consideration there.

The backdrop to Brindley's rise to national prominence lay in a period of relative security and economic wellbeing in the country. The Catholic Insurrection of the second Jacobite rebellion had been put down. Trade was opening up to new colonies. A growing population, fuelled by the agricultural revolution and employed by the growing

Brindley's Mill, Leek.

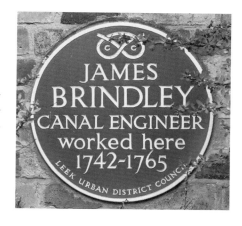

industrialisation of the country, led to a
demand for improved transportation of goods
within Britain. These events were stimulated
by low interest rates that enabled businesses
across the country to grow rapidly.

In 1752 during the course of his many
and varied engineering projects in the
Midlands and North-West England, Brindley
combined both aspects of mechanical and
civil engineering. He introduced an ingenious method of pumping water from Wet
Earth Colliery near Manchester for John Heathcote. The system involved the building
of a siphon under the River Irwell that fed water to an underground waterwheel. This
wheel, in turn, drove pumps that exhausted water from the mine: a perfect ecological
arrangement.

Continuing his mechanical engineering, Brindley was also involved in building steam
engines. In 1756 he had observed the working of a Newcomen engine at Bedworth, and
in 1758 he patented a float arrangement in the boilers which kept the water at a constant
level (Patent 730). The engine was for Mr Thomas Broad of Vivian Fenton and the
float arrangement was possible given the very low steam pressures of early atmospheric
engines. The device was used for many years afterwards by various engineers including
James Watt. Mill work continued too, and, as Christine Richardson has estimated, by 1758
James Brindley had approximately twenty men working for him on a freelance basis.
From this team, Brindley recruited Samuel Simcock once the canal era commenced. In
1758 Brindley also took the first step towards the work for which he is nationally famous
when, with his future father-in-law John Henshall, and John Smeaton, he surveyed a
canal to link Burslem to Burton with a branch to Bristol. This was the genesis of the
Great Cross that was to link the major rivers across the watershed of England and the
ports of London, Liverpool, Bristol and Hull; thus providing the template of the English
canal system.

The impetus for such a scheme lay in the growing number of industrial towns of
the North West and Midlands, and the development of the agricultural hinterland
required to feed the growing population of these towns. This work, and the building
of a water engine in 1759 at Cheadle, brought Brindley to the attention of the Duke
of Bridgewater's agent, John Gilbert. The Duke employed Brindley on his canal from
Worsley to Manchester in 1759 and from that date Brindley's 'unusual talents unfolded
to their full extent'.

Although the Newry Canal in Ireland, a summit canal 18 miles in length completed
by Thomas Steers in 1742, and the Sankey Cut, a lateral canal completed by Henry Berry
in 1757, had preceded the Bridgewater Canal, it was the latter canal that marked the
turning point in the history of the construction of canals in Britain. The reason for this
was that the Bridgewater demonstrated dramatically how effective a well-organised and

James Brindley's Day Book. Record of journey to London from Manchester. (By kind permission of ICE)

executed canal could be in transporting heavy goods and reducing the price of valuable commodities such as coal. The talismanic Barton Aqueduct was also regarded as a marvel of the modern age, and people visited to see boats travel across the River Irwell 39ft in the air. John Phillips observed that in it, 'grandeur, elegance and economy are happily united'.

At this juncture it is worth remembering just how bad some of the roads were in the late eighteenth century, despite the growth of turnpike roads, and why canals were needed so desperately. Wedgwood recorded that 'pack horses were heavily laden with coal and tubs of ground flint from the mills, and hacked to pieces by whips and cruel drivers whilst floundering knee deep through the muddy holes and ruts that were all but impassable'.

Francis Egerton, the Third Duke of Bridgewater, has had his story told elegantly and in great detail by Hugh Malet. From being an unwanted, sickly member of the aristocracy he suffered an abusive childhood at the hands of his stepfather and then became the rejected suitor of one of the great beauties of the age, Elizabeth Gunning. Eventually, the seemingly backward duke found inspiration while on a Grand Tour of the Continent, when he saw the Languedoc Canal, the Canal du Midi, in France and became fixated on creating his own canal on his northern estates – from his mines at Worsley to the centre of Manchester. From a wastrel figure to public benefactor, the story has all the ingredients of sensational fiction and was even stranger in that it was true.

As mentioned, on the Bridgewater Canal Brindley built the substantial and innovative aqueduct at Barton, over the River Irwell. The aqueduct was 200yds long, 12yds wide

The Duke of Bridgewater posing in front of
Brindley's Barton Aqueduct.

and 39ft above the river with a central
arch spanning 63ft. He also assisted John
Gilbert in creating the underground
tunnels at Worsley from which coal was
mined and placed directly on to vessels
for transportation to Manchester. Work by
historians Lead and Malet shows that Gilbert
was the chief instigator of the underground
mines, but Brindley was involved in their
construction; providing ventilation through
water-driven bellows as well as building
portable cranes and inclined planes linking
the different canal levels within the mine.
The underground coal mine leading directly
to the canal was a practice replicated later by
Brindley at Harecastle Tunnel on the Grand
Trunk Canal.

Apart from capturing the public's
imagination, the canal established Brindley
in the minds of future canal proprietors as
the man to consult on engineering. In fact,
what had been important was Brindley's
re-survey of the original Bridgewater Canal
across Barton Aqueduct, shortening the distance into Manchester. His survey of a second
line to Preston Brook was also significant. Here the new canal would join with the
planned the Grand Trunk Canal and, via the Great Cross of canals, to every point in the
country. Thirdly, it was the first canal, according to Rees, where puddle lining of the canal
bed was used systematically throughout. By 1800 Manchester, due largely to the work of
the duke, John Gilbert and James Brindley, had become the second-largest city in Britain,
with a population of 84,200 inhabitants.

The building of this first canal has been described in detail by historians from the
eighteenth century through to our own. We hear of the work of the Triumvirate: the
duke; John Gilbert, the duke's agent; and James Brindley himself. Accounts abound of
Brindley's progress to Parliament, his successful oratory and his 'ocular demonstrations'
in the Chamber of the House with puddle, the waterproof sealant that lined the canal
bed. He mixed up puddle in the chamber itself so that their lordships could see, and
understand, what was being proposed by way of construction. Intimate details also exist
from Brindley's Day Books of his journey, and its cost, from Manchester to London: how
his horse fell over in the frost at Coventry; how much money he spent on clothes and
shoes before attending Parliament.

The issue under examination, however, is how he utilised other people's skills to
promote this project and those that were to follow. Samuel Simcock, foreman, carpenter

and brother-in-law, had been brought to assist with the survey and construction of the first Bridgewater Canal in 1759. He was sent to survey the Sankey Brook and its locks, presumably as a dry run to Brindley's work. Hugh Henshall was then recruited to survey the second branch of the Bridgewater Canal to Preston Brook, completed in 1776. Brindley was forming a team that he would use to realise his, and others', plans of a canal network; not only linking the major rivers of England but extending the canal system further. James Brindley had a good eye for country and would undertake an 'ochilor servey' or 'recconitoring', a visual survey of the lie of the land with no instruments to measure levels. From here, however, he left the detailed surveying to assistants who were usually established land surveyors such as Hugh Henshall and Robert Whitworth.

The second line of the Bridgewater Canal, authorised in 1762, was linked to the original canal at Trafford and ended near Runcorn on the Mersey. On this canal, Brindley constructed major embankments at Bollin and Stretton. In these he devised the use of a caisson of oak and deal planks whilst the earth was being built up in support. The success of the two Bridgewater canals led the Duke of Bridgewater and Josiah Wedgwood to propose the building of the Trent and Mersey Canal, to stimulate the growth of industry that was beginning to develop in the Midlands and North West. Brindley called the canal 'The Grand Trunk', in allusion to 'the main artery of the body from which branches are sent off for the nourishment of distant parts'. Efforts to secure permission for building The Grand Trunk were assisted by the relationship between the Duke of Bridgewater and Earl Gower, whose estates in Staffordshire lay across the proposed route of the new canal. James Brindley, with the financial and political support of a brace of dukes plus a major captain of industry, Josiah Wedgwood, was in prime position to undertake this work following his success with the construction of the Bridgewater Canal. According to Rees,

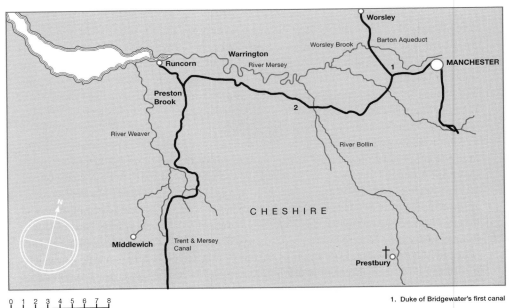

Fig.1: The Duke of Bridgewater's two canals.

Day Book. Brindley's shopping list for clothes prior to attending Parliament, 1762. (By kind permission of ICE)

'Mr James Brindley owed much of his well earned fame to the happy contrivance and complete execution which he displayed in every part of this great concern.'

In 1760 James Brindley entered into a partnership with Thomas Gilbert, his brother John Brindley and Hugh Henshall to purchase the Turnhurst Estate and the Golden Hill Colliery which lay below it. Brindley invested £543 8d. In the same year, Brindley also became a partner with his brother John, a potter, in the Longport Pottery. In 1761 Brindley moved north from Leek permanently, to his rented workshop owned by John Brindley. It was adjacent to Josiah Wedgwood's home in Burslem. However, the reality was that home for Brindley was in coaching inns and lodgings near his work.

The Treaty of Paris, signed in 1763, brought to an end the Seven Years War against France – which had been fought mainly in America. This resulted in the further expansion of the British Empire in colonies taken from France, and the opportunities for capital to be directed towards much needed public works and away from military production. Clive's victory against the French at Plassey in 1757 had given the British dominance in India, and Wolfe's at Quebec, in 1759, dominion of Canada. In the second half of the eighteenth century Britain then entered the first Industrial Revolution and Brindley's canals were there as 'vital midwives' to that birth.

In 1765 Brindley, Thomas Gilbert, Josiah Wedgwood and John and Hugh Henshall met at Wedgwood's house in Burslem to discuss plans for building a windmill for crushing flint and improving access roads. They also discussed the survey of the canal navigation that was to become the Grand Trunk Canal. Widespread canvassing and political lobbying eventually led to a meeting at the Wolseley Bridge Inn in Staffordshire on 30 December 1765, with Earl Gower the chair. Wedgwood, Brindley and other would-be investors were in attendance to gain local support for proceeding with their plans for the great canal to Parliament. Meanwhile, there was a rare break from canal and engineering activities for James Brindley when he married Anne Henshall, sister of Hugh and daughter of the land surveyor John Henshall, at Wolstanton church on 8 December 1765. The witnesses at the wedding were Hugh Henshall and John Mills, with the bride nearly thirty years younger than the groom. Initially, they lived together with the Henshall family at Bent Farm near Newchapel, before moving to Turnhurst Hall on the top of Harecastle Hill.

Above: Wolseley Arms, near Little Haywood, where the canal committee met to plan the Trent and Mersey Canal.

Left: Plaque, Wolseley Arms.

On 14 May 1766 the Trent and Mersey Canal became, by Act of Parliament, a joint stock company. This innovation gave the canal company the right to compulsorily purchase land which was needed for construction of its canal. It also conferred the right to levy tolls on goods carried. The canal, unlike the Bridgewater Canal which had been financed by one wealthy aristocratic landowner, was financed by businessmen who operated along its route, particularly Josiah Wedgwood and the potters in Stoke. This was to set a precedent for future canal and railway construction. The canal was designed to link the Duke of Bridgewater's canal near the Mersey at Preston Brook, to the River Trent at Wilden Ferry, south of Derby. Clearly James Brindley had vested interests in the project, and he became consulting engineer to the new canal. Hugh Henshall became the resident engineer and John Sparrow, a solicitor from Newcastle-under-Lyme, the clerk to the canal company. James' brother John was a member of the Trent and Mersey Canal committee, as was Matthew Boulton.

The building of the canal, over 93 miles long with seventy-six locks, 160 aqueducts, 213 road bridges and six branches, was at the heart of the Great Cross; the aorta of the Brindley system. Of all aspects of engineering, however, it was Harecastle Tunnel, north of Stoke-on-Trent, that caught the public's imagination. Erasmus Darwin called it 'the astonishment of ages'. John Smeaton, in his joint survey with Brindley and Henshall six years earlier, had originally recommended a deep cutting be constructed at Harecastle, but James Brindley had persuaded the company that a tunnel would serve their interests better.

Harecastle was, in effect, the first substantial canal tunnel for navigation in England; the two tunnels previously constructed on the system were very short, whereas the Worsley tunnels were in reality navigable drift mines in pursuit of coal, albeit very complicated ones. The tunnel at Harecastle, however, was 2,880yds in length, 9ft wide and 12ft high. Brindley surveyed the line over Harecastle Hill and from this sank fifteen vertical shafts from which headings were driven on the line of the tunnel. The tunnel had to be dead level and straight. It is unclear as to whether Josiah Clowes assisted Brindley with this work, but it is possible he did, having experience of local mining and having reported to a parliamentary select committee that he had worked with Brindley. Certainly, Hugh Henshall would have been available to assist as resident engineer, as he owned land with Brindley on top of the great tunnel. He was later left, after Brindley's death, to complete the work.

The ground was difficult to tunnel: millstone grit plus soft earth and a great deal of water was encountered. Brindley had to build wind pumps and employ Newcomen steam engines to pump out the water. He also installed stoves at the bottom of tunnel

Harecastle Tunnel, north portal.

pits to ventilate the mine, as well as incorporating cross galleries in the tunnel to excavate coal. Like the aqueduct at Barton, the tunnel was an extremely imaginative piece of engineering for its day as nothing on this scale had been attempted before. Brindley was over-optimistic about the time it would take to build, but his scheme eventually came to a successful conclusion under Hugh Henshall's stewardship. A.R. Saul tells us that a small dry tunnel communicated with the colliery tunnel branching from the main Harecastle Tunnel; access was behind Latebrook House and it was used in the Great War as an air-raid shelter.

Whilst the Grand Trunk was under construction, James Brindley assisted with the building of the Droitwich Canal, although it was left to Robert Whitworth to undertake the survey and John Priddey to engineer the canal. Brindley also visited Scotland to give advice on the Forth and Clyde Canal, as well as undertaking a survey of the Chesterfield Canal. The pressure on Brindley of growing ill health (he was suffering from diabetes) led James and Anne to take a brief break in Buxton, near his original home, in the summer of 1767. This was too short a rest to help his weakening constitution, but despite the growing effects of his illness James Brindley turned his attention again to the construction of the burgeoning canal network, on which he worked for another five years.

Brindley had surveyed the Staffordshire and Worcestershire Canal a year earlier in 1766 and continued to visit on a regular basis, as his other commitments would permit, to oversee the works there. Indeed, this branch of the Great Cross was completed several years before the Trent and Mersey Canal. The final arm of the Cross involved the building of the Coventry and Oxford canals, started in 1768. These were to link the Midlands and the North to the Thames and London. Both these canals were surveyed initially by

Weir, Stewponey Lock, Staffs and Worcs Canal.

Fradley Junction, at the meeting point of the Trent and Mersey Canal and the Coventry Canal: the route to the South.

Brindley, although they were to be subsequently re-surveyed and re-routed by Whitworth and Simcock. Here, Brindley's vision and organisational ability of others came to the fore. Indeed it was as well James Brindley established his school in the way he did, as he died with all his canals unfinished except the Staffordshire and Worcestershire Canal and the Droitwich Canal.

It was clearly a well-planned move. Brindley acquired a team of men to survey and take levels. He also recruited chain and stake holders as well as competent draughtsmen to produce detailed maps and, if possible, engineers with a knowledge of hydraulics. Josiah Clowes became, under Brindley's supervision, an expert in tunnel construction. Robert Whitworth and Hugh Henshall complemented his surveys with detailed and excellent maps. Thomas Dadford became a specialist in lock construction. Throughout, his reliable brother-in-law and fellow millwright, Samuel Simcock, worked as surveyor or resident engineer on all arms of the Great Cross. In the nineteenth century, such apprenticeship became more organised and formal. For example, during the age of railway construction, civil engineers such as William and Peter Barlow were apprenticed to Henry Robinson Palmer on various building projects before moving to undertake work in their own right. Despite the non-existence of such a clear apprenticeship scheme in his day, James Brindley plainly saw the need to develop a collegiate system to train the new engineers in the emerging hydraulic technology.

Brindley's method of tutoring his engineers is exemplified by his use of 'workcamps' and the mobility of personnel. He moved Joseph Parker from the Coventry and Oxford canals to spend a month under Hugh Henshall's tutelage on the Staffordshire and Worcestershire Canal. For this work, Henshall was paid 10 guineas. Shortly afterwards, Christine Richardson records, John Bushel, the contractor for the locks on the Droitwich Canal, arrived for the same purpose. She comments:

Elsewhere, the senior men were moved around to maximise the benefit of their expertise. Samuel Simcock … had been involved in the Staffordshire and Worcestershire Canal … was at the Eastern end of the Trent and Mersey, before being allocated to the Birmingham Canal as senior engineer.

In 1768–69 James and Anne took up residence in Turnhurst Hall. The Hall, unlike the surrounding lands, was owned by the Bowyer/Egerton and Alsager families. Handel stayed there but, unfortunately, the Hall was pulled down in 1929. The front door of the Hall was saved and re-erected at the entrance to the school which adjoined St Michael's church in Chell. Brindley is reputed to have worked in a small tower in the corner of the grounds – Samuel Smiles described it as a summerhouse. Anne was in charge of the paperwork, as can be seen from the letters sent by Brindley to various canal companies bearing his signature, but the letter scribed in a different hand. James was everywhere but Turnhurst, surveying and engineering canals in the North and North East as well as the Midlands and those canals that made up the Great Cross. Over time the couple had two children, Anne and Susannah.

It is a favourite tale that James Brindley made a canal in the garden at Turnhurst equipped with a model of a working lock. If so this would have been one of the earliest examples of model-making before full-size design and construction. Klemperer and Sillitoe, in their 1995 account of archaeological excavations at Turnhurst, conclude that although there was a canal in the grounds, which dated in part from the eighteenth century, there is no direct evidence to suggest that it contained Brindley's trial lock; rather it was an ornamental water feature. However, the dig was not completed and therefore the true situation is unknown. It is possible that the oral tradition is true and has just not been confirmed by first-hand evidence. Certainly Brindley would have needed to

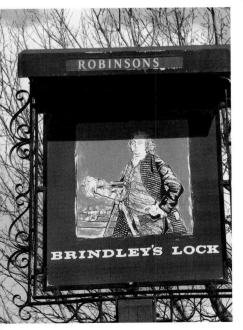

know about lock construction in more detail as he had limited, if any, experience of lock construction at the start of building the Trent and Mersey Canal. Peter Cross-Rudkin says it is uncertain whether Brindley constructed locks on the Halifax branch of the Calder and Hebble Navigation in 1765. What was evident was that Brindley standardised the length and breadth of locks on the Trent and Mersey Canal: 74ft 9in long by 7ft wide.

Wedgwood regularly visited James and Anne Brindley at Turnhurst Hall. In 1768 Wedgwood brought Mrs Brindley a handsome mahogany tray with accompanying tea service. Mrs Brindley responded with a visit to Etruria to see the factory and how her husband's canal was progressing in front of the works. As she was often on her own, with James out in the field surveying and engineering,

The link to Brindley lives on in the name of a public house now built on the site of Turnhurst Hall, Stoke.

she became good friends with the Wedgwoods and they met often. Brindley would bring Josiah fossils extracted from the tunnel for examination. As Wedgwood once wrote: 'Mr Brindley and his lady called here on the way home this morning – we are to spend tomorrow with them at Newchapel, and as I always edify full as much in that man's company as at Church, I promise myself to be much wiser the day following.' Josiah even passed on one of the fossils of an elephant or mammoth bone to his friend from the Lunar Society, Erasmus Darwin. From these and other fossilised remains, Darwin arrived at what is now called the 'theory of common descent'; the belief that all life as we can see it today is descended from one microscopic ancestor, 'a single living filament'. It is interesting to reflect that apart from being the country's first significant canal tunnel, Harecastle also had a part in engendering the then revolutionary scientific thought of evolution.

The majority of insights into James Brindley's character and unique method of working are afforded by Hugh Henshall. The revelation into Brindley's ingenuity and thinking process were recorded by Henshall in an article he published in the *Universal Magazine* in 1780. Henshall describes how Brindley invented a special machine for making tooth and pinion work reliably on gear wheels. Secondly, he outlined how the upshaft fires, set in Harecastle Tunnel, increased the amount of fresh air to the workings. Finally, we hear from Henshall about a machine, devised by Brindley at Manchester, for landing coal at the end of the Bridgewater Canal. Brindley designed an undershot wheel to provide power to the hoist at the Castlefield warehouse. Henshall concludes that 'this pleasing idea' represented Brindley's ambition of 'diminishing labour by mechanical contrivance'. An interesting notion in one whose death was accelerated by overwork.

Henshall's accounts of his brother-in-law's character and thinking processes were subsequently recorded, with Josiah Wedgwood's observations, for Dr Kippis to include in his *Biographia Britannica*. This information was then drawn upon by Samuel Smiles when he wrote his biography of Brindley. We learn that Brindley would retire to bed for two

Harecastle Tunnel, south portal.

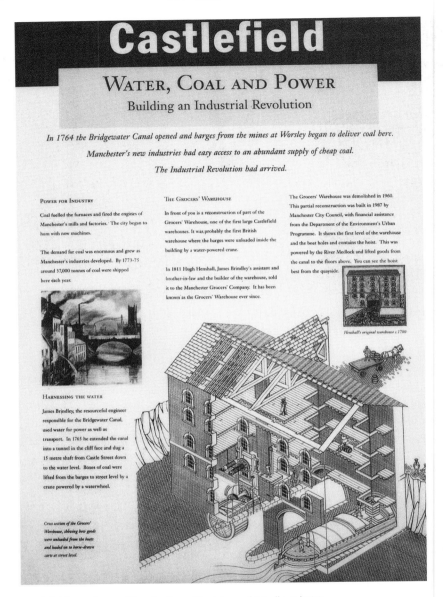

Castlefield

WATER, COAL AND POWER
Building an Industrial Revolution

In 1764 the Bridgewater Canal opened and barges from the mines at Worsley began to deliver coal here.
Manchester's new industries had easy access to an abundant supply of cheap coal.
The Industrial Revolution had arrived.

POWER FOR INDUSTRY

Coal fuelled the furnaces and fired the engines of Manchester's mills and factories. The city began to hum with new machines.

The demand for coal was enormous and grew as Manchester's industries developed. By 1773-75 around 37,000 tonnes of coal were shipped here each year.

THE GROCERS' WAREHOUSE

In front of you is a reconstruction of part of the Grocers' Warehouse, one of the first large Castlefield warehouses. It was probably the first British warehouse where the barges were unloaded inside the building by a water-powered crane.

In 1811 Hugh Henshall, James Brindley's assistant and brother-in-law and the builder of the warehouse, sold it to the Manchester Grocers' Company. It has been known as the Grocers' Warehouse ever since.

The Grocers' Warehouse was demolished in 1960. This partial reconstruction was built in 1987 by Manchester City Council, with financial assistance from the Department of the Environment's Urban Programme. It shows the first level of the warehouse and the boat holes and contains the hoist. This was powered by the River Medlock and lifted goods from the canal to the floors above. You can see the hoist best from the quayside.

Henshall's original warehouse c.1780

HARNESSING THE WATER

James Brindley, the resourceful engineer responsible for the Bridgewater Canal, used water for power as well as transport. In 1765 he extended the canal into a tunnel in the cliff face and dug a 15 metre shaft from Castle Street down to the water level. Boxes of coal were lifted from the barges to street level by a crane powered by a waterwheel.

Cross section of the Grocers' Warehouse, showing how goods were unloaded from the boats and loaded on to horse-drawn carts at street level.

Castlefield warehouse, Manchester, with picture of Brindley's hoist.

or three days with problems and then emerge, like Archimedes, with the correct solution. We also gain insights into the effect of the theatre in London upon James when John Gilbert and his wife, Lydia, took the engineer on 17 January 1762 to see David Garrick's performance in *Richard III*. The experience completely overwhelmed Brindley and he retired to bed for several days in shock, the play having 'disturbed his ideas and rendered him unfit for business'.

Wheel from Brindley's hoisting machinery, Castlefield, Manchester.

John Philipps said of Brindley: 'He never seemed in his element, if he was not either planning or executing some great work, or conversing with his friends upon subjects of importance.' He added that Brindley had 'no common diversions of life'. Stories abound: Brindley describing 'water as a giant, safe only when laid upon its back'; his demonstrations to the Select Committee of the House of Lords with his carved model aqueduct constructed from cheese – another 'ocular demonstration'; also Brindley writing in his Day Book about making a good cloth, or rupture plaster, to combat back pain; plus the efficacy of drinking one's urine to improve health, but only when it was warm.

The combination of hard work, drinking, gourmandising, self-administered medical treatment and a suspected sweet tooth (he courted Anne Henshall by taking her gingerbread) eventually killed Brindley. Diagnosed eight years prior to his death as a diabetic by Erasmus Darwin, Brindley clearly lacked robustness in his sixth decade and caught a chill whilst surveying the Caldon Canal in the late summer of 1772. Put into a damp bed at the nearby Red Lion Inn at Ipstones, Brindley's condition deteriorated and he was eventually moved to Turnhurst Hall where, on 25 September 1772, he died of pneumonia, leaving the mantle of canal engineering on his brother-in-law Hugh Henshall and the other members of his school of engineers.

Brindley died intestate with an estate valued at approximately £7,000. Hugh Henshall became ward to his children and managed his widow's affairs for a period, attempting to recoup monies owed by the Duke of Bridgewater and the Gilberts to the Brindley family. In 1770 Brindley had had his portrait painted in London by Francis Parsons but had refused to pay the £60 for the finished work saying it was too expensive. On his deathbed he relented and instructed Henshall that the portrait be bought for his wife. It now hangs in the National Portrait Gallery in London. In Brindley's obituary is the

Erasmus Darwin's house at Lichfield. Darwin was Brindley's doctor.

passage: 'Mr Brindley had long been sensible of the precarious situation with his health, and wished to be proceeded in his profession by his brother-in-law Mr Henshall … he spared no pains to qualify him for that important trust …'

The verdict Dr Erasmus Darwin passed on James Brindley was that he 'was better qualified to be a contriver rather than a manager of great design', but with Benjamin Franklin he was one of only two men Darwin referred to as 'immortal'; praise indeed. Brindley had drawn about him by his wit, intelligence and force of personality, a school of skilled engineers who were able to complete his life's work. T. Lowndes, in his *History of Inland Navigation in Lancashire and Cheshire*, says 'sensible that he one day will cease to be, Brindley selects men of genius teaches them the power of mechanics and employs them in carrying out the serious undertakings in which he is engaged'. This is ultimately the strategy referred to in this book as the 'School of Engineering'.

Erasmus Darwin was grief-stricken by Brindley's death. He wrote to Wedgwood: 'I have always esteem'd him to be a great Genius, and whose loss is truly a public one. I don't believe he has left his equal.' Darwin thought that this genius should have been recognised by a grateful nation, and that a national monument to the great engineer should have been erected in Westminster Abbey. In this matter, the grandfather of Charles Darwin and human evolution was undoubtedly correct. James Brindley was, however, buried at St James church, Newchapel; a far cry from Westminster Abbey. In the chapel a tablet was erected which reads:

This new church of St. James, Newchapel, replacing one ruined by mines was opened for Divine Service, February 14th, 1880. The cost was in part subscribed in memory of

James Brindley's
tomb, St James
church, Newchapel.

James Brindley who died at Turnhurst in this parish Sept 25th, 1772, and was buried in the churchyard adjoining.

A plaque was added to Brindley's tomb in 1958 by The Staffordshire Antiquarian Society in recognition of his achievements. A few feet away from James lie his wife Anne and his brother-in-law Hugh Henshall, who, although not his equal in engineering, was well trusted to carry on Brindley's life's work.

Cyril Boucher records that James Brindley had a natural son by Mary Bennett, who he met whilst working in Manchester. The boy was baptised John at Burslem on 31 August 1760. It is claimed that through this union, James Brindley became the great-great-grandfather of the novelist Arnold Bennett.

2

HUGH HENSHALL 1734–1816

Mr Brindley had 'a peculiar regard for him of whose integrity and abilities in conducting these works he had the highest opinion'

Brindley's obituary, *Aris's Birmingham Gazette*

Hugh Henshall, surveyor, engineer, canal carrier, canal proprietor and pottery manager, was born in 1734 in north Staffordshire, probably at Wolstanton. His parents were John Henshall, also a land surveyor, and Anne Cartwright, who had married in Wolstanton on 6 January 1730. The Henshall family had lived in the parish of Wolstanton, north of Stoke-on-Trent, for a hundred years before the birth of Hugh. At the time of Hugh's birth they lived in Newchapel. Hugh became a crucial partner in the Brindley school of engineering.

John and Anne Henshall had five surviving children, who were born between 1731 and 1747. In January 1731 the eldest child Jane Henshall was born, and Hugh was born in 1734, the same year that his older brother John died in infancy. On 1 January 1749/50, Jane Henshall married William Clowes at Norton-in-the-Moors. Clowes was a local landowner with strong coal mining interests. He was also the elder brother of Josiah Clowes, the early canal and tunnel engineer who was later to become part of the Brindley school of engineering and would form a close friendship and working partnership with Hugh Henshall and Robert Whitworth. In 1747 Hugh's youngest sister Anne was born and, in 1765, she married James Brindley at Wolstanton. Hugh was a witness at the wedding.

We know from Eliza Metyard's biography of Josiah Wedgwood that Hugh Henshall went to school with Josiah Wedgwood at Newcastle-under-Lyme, a 7-mile journey for the young Hugh every day. The school was run by John and Thomas Blunt and was housed in an old, half-timbered building in the market place. The tuition was provided on two floors and, unusually for the time, it also provided education for girls. Indeed, several years after Hugh had left his younger sister Anne attended and was taught by Mrs Blunt. She was taught English and Mathematics as well as practical subjects. This was to stand her future husband, James Brindley, in good stead as Anne could and did write letters for him as well as keep the books. Hugh's abilities as a surveyor were inherited from and developed by his father, John Henshall, who was a land surveyor and had worked with James Brindley on the early Trent and Mersey Canal surveys.

As a young man Hugh Henshall worked as a surveyor in north Staffordshire, Cheshire and Gloucestershire. In 1717, before Hugh was born, a proposal was made by Thomas

THE HENSHALL FAMILY

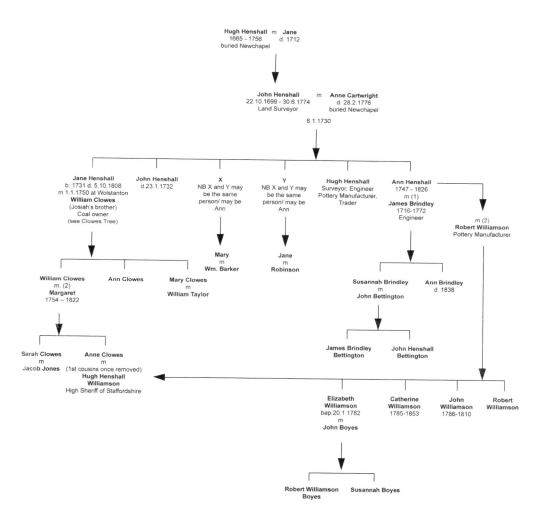

Congreve to link the rivers of the Trent and Mersey. In 1758 Henshall met James Brindley and, with John Smeaton, made a survey of such a navigable waterway. Local records reveal Henshall's work on Staffordshire estates with a view to this enterprise. On 30 January 1760 Mr Lingard, later to be an employee of the Trent and Mersey Canal, was writing to Hugh Henshall regarding the progress of preliminary digging at Findon. In the letter to Henshall, Lingard mentions the discovery of clay appropriate for the manufacture of bricks and talks of a proposed meeting with Samuel Simcock, Brindley's other brother-in-law and fellow engineer, at the Plum Pudding Public House, Findon: 'a favourite watering hole'. Alongside this work, Hugh Henshall also had time for social occasions. On 30 December 1762 Hugh was best man at Josiah Clowes' marriage to Elizabeth Bagnall, at Norton-in-the-Moors church. Also in this year, Henshall assisted with the survey and mapping of the second Bridgewater Canal, where it is almost certain he met and worked with Samuel Simcock.

In 1765 Hugh Henshall was involved with Robert Pownall on a survey of the River Weaver from Winsford to Lowton via Middlewich and Nantwich. This work must have been of great value to Henshall a decade later when he came to be engineer on the northern section of the Trent and Mersey Canal, including the Saltersford and Barnton tunnels, after Brindley's death. In the following year in May, Hugh Henshall was also involved in a survey of the River Severn. It is clear that the various navigation links that became part of the Great Cross were already fomenting in the minds of Brindley and Josiah Wedgwood. Given his previous work, Hugh Henshall must have been consulted as one of the surveyors. As Michael Chrimes reveals, during the building of a canal 'an engineer might engage the services of a local surveyor to assist in the survey of the line and land valuation'.

Henshall's relationship with James Brindley was personal as well as professional. From the account in Smiles' *Lives of the Engineers*, we know that Brindley, at the age of forty-nine, fell in love with Hugh Henshall's youngest sister Anne, who was eighteen. He brought Anne gingerbread and paid court to her whilst she was still a schoolgirl. James Brindley and Anne Henshall married on 8 December 1765 at St Margaret's church, Wolstanton. Hugh Henshall and John Mills were the witnesses. James may well have gone to live with the Henshall family for a while with his new wife after the wedding, as Josiah Wedgwood wrote that he was planning to visit the newly married couple on 23 February 1765 at Newchapel.

On 14 May 1766 the Trent and Mersey Canal, or the Grand Trunk Canal as it was then called, became, by Act of Parliament, a joint stock company. This innovation for funding the building of canals gave the company the right to compulsorily purchase the land which was needed for construction of its canal. It also conferred the right to levy tolls or goods carried along the canal. The canal was funded by shares that could be bought and sold. (This purchase and sale was regulated by Parliament in 1720 after the South Sea Bubble deception.) Unlike the Bridgewater Canal which was financed by one wealthy aristocratic landowner, this new canal was financed by businessmen who operated along its route, particularly Josiah Wedgwood and the potters in Stoke. This was to set a precedent for future canal and railway construction. The Grand Trunk was designed to link the Duke of Bridgewater's canal near the Mersey at Preston Brook to the River Trent at Wilden Ferry. Clearly, Hugh Henshall had vested interests in the project.

Hugh Henshall drew the parliamentary map of the Trent and Mersey Canal, as built for James Brindley, and on 3 May 1766 he was appointed clerk of works to what was then the premier engineering project in the country. His annual salary was £150 for himself and an assistant. Initial work commenced westwards from Wilden Ferry towards Stone. In the same year Henshall was surveying roads in Newcastle-under-Lyme. It is certain that these roads were linked to the engineering and construction of the Trent and Mersey Canal. The Trent and Mersey Canal was 12ft wide at the base. It had a depth of 3ft and could accommodate boats 70ft long and 6ft wide. The boats drew 2ft 6in when carrying a full complement of 20 tons. On 26 July Josiah Wedgwood cut the first sod and James Brindley wheeled it away in a barrow. The inaugural celebration was attended by Hugh Henshall, Josiah Clowes and 'many respectable persons of the neighbourhood, who each cut a sod to felicitate the work'.

In 1767 Hugh Henshall, as clerk of works and 'ingenious and practical assistant' to James Brindley, was involved in the purchase of land for the new canal. On 27 January we hear of a purchase from Mr Lawton, signed by the solicitor to the Trent and Mersey Canal Company, Mr John Sparrow. A month later land was purchased at Kings Bromley from George and John Whiston, Sarah Barker and William Cross for the building of the canal. The agreement was signed by James Brindley, Hugh Henshall and William Cross. In December 1767 the canal was laid out in front of Etruria Hall, Josiah Wedgwood's property. This part of work gave insight into Brindley's irascibility as Wedgwood and Henshall were anxious to abide by Brindley's instructions, or 'Mr Brindley would go mad'. Josiah Wedgwood teased Henshall over the issue by calling Henshall 'an inflexible vandal'. However once the building was started, the revised line of the canal near Etruria did cause some difficulties between John Brindley, Josiah Wedgwood and Brindley himself. John, as a potter in Burslem, felt that Wedgwood, through Brindley's and Henshall's survey, had gained an unfair trading advantage by the canal being laid out in front of Wedgwood's new works. This difficulty was eventually overcome by independent arbitration and it was agreed that the canal *was* following the best route for engineering purposes.

On 25 March 1768, Hugh Henshall was negotiating with the purchase of land from Lord Paget. On this date the address on his correspondence indicates clearly that Henshall was still living at Newchapel, very possibly with his elderly parents and grandparents. Occasionally, however, he took a break from canal work. In 1771 we hear of Henshall surveying and drawing a plan of the manor and parish of Norton-in-the-Moors, Stafford, that belonged to Charles Bowyer Adderley. Henshall's beautiful map can still be seen in the Potteries Museum and Art Gallery, Hanley.

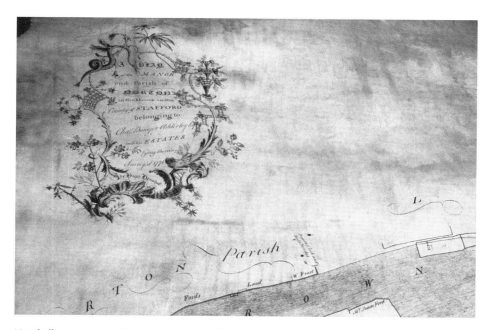

Henshall's signature on the map he drew for Charles Bowyer Adderley. (By kind permission of the Potteries Museum and Art Gallery, Hanley)

The watershed of the new canal lay on the high ground near Ranscliff, north-west of the Potteries, which Hugh Henshall, James Brindley, John Gilbert and Josiah Clowes owned. The canal tunnel at Harecastle, when built, passed through the coal, ironstone and millstone grit under much of their land. The quartet of families clearly had interests in mining. At 2,880yds the mammoth tunnel, which had subsidiary tunnels branching from it to coal measures, reflected the practice at Worsley. Hugh Henshall was left, after Brindley's death in 1772, to complete the task of engineering the great tunnel and the difficult northern section of the canal. The Trent and Mersey Canal descended into the Cheshire plain after navigating the tunnel at Harecastle, and formed a junction with the Duke of Bridgewater's canal at Preston Brook near Runcorn. Hugh Henshall was also responsible for the cutting of this second Bridgewater Canal through to Runcorn, thus finally completing the link between the rivers Trent and Mersey. Henshall also had to complete several aqueducts on the Trent and Mersey Canal, including the twelve-arch aqueduct over the River Dove which was completed in 1770.

In 1768 James Brindley surveyed the Staffordshire and Worcestershire Canal, designed to link the Grand Trunk at Great Haywood to the River Severn at Stourport. Hugh Henshall assisted with part of this survey. In September 1766 the joint stock company paid Henshall £42 19s 1d. This was almost certainly for drawing Brindley's maps of the undertaking. The final part of James Brindley's Great Cross came in 1768–69, when the Coventry and Oxford canals were planned to link the Trent and Mersey Canal at Fradley to the Thames at Oxford. The Oxford Canal was completed in 1790 by Samuel Simcock. The Oxford Canal also was surveyed by another of Brindley's school, Robert Whitworth. Although Whitworth was to become a more prominent figure in the field of civil engineering than Henshall, it was Henshall that Brindley entrusted with engineering the completion of the Grand Trunk Canal. Indeed, it is very likely that the actual building of the canal fell to Henshall, with Josiah Clowes entrusted with work on Harecastle

Dove Aqueduct, near Burton, Trent and Mersey Canal.

Norwood Tunnel on the summit of the Chesterfield Canal.

Tunnel. There is sufficient evidence – given Brindley's regular absences from the works, the length of canal Henshall surveyed and built, Clowes' knowledge of mining and the number of years that Henshall worked on the canal – to guess this might have been the case.

On 12 February 1768 Brindley asked Henshall to survey the Birmingham Canal, which branched from the Staffordshire and Worcestershire Canal at Aldersley Junction near Wolverhampton. This vital canal set in place a crucial link with the Trent and Mersey Canal, via the Staffordshire and Worcestershire Canal, and the growing city of Birmingham and the neighbouring Black Country. In the event, Henshall did not attend and Samuel Simcock undertook the work with James Brindley, assisted by Robert Whitworth.

Anne and James Brindley moved to Turnhurst Hall in either 1768 or 1769. The Hall was situated on the top of Golden Hill over the developing canal. This estate was owned in part by the Henshall and Gilbert families. Hugh Henshall never married but took a supportive interest in his family's affairs, as described later. He enjoyed close family ties with his sister Anne and his many nephews and nieces from her two marriages, and also those from his elder sister, Jane Clowes. He seems to have adopted an avuncular role throughout his middle age and later life as he was there to support and protect the interests of his family, as well as pursue his civil engineering and business interests. On 12 April 1771 Henshall was admitted into the Smeatonian Society, which was at this stage simply called the Society of Civil Engineers. There he met Robert Whitworth and Thomas Dadford; at his first meeting Thomas Yeoman was in the chair.

On Tuesday 13 June 1769, a particularly hot day, the Brindleys and the Henshalls attended the official ceremony, by the banks of the Grand Trunk Canal, to open Josiah

Wedgwood's new factory at Etruria. Josiah gave a brief speech and then, in the spirit of good publicity, he turned the clay on a special wheel into Etruscan vases while Thomas Bentley cranked the handle and the crowds watched in wonder. On the black basalt vases was inscribed: 'The arts of Etruria are reborn.'

A few years later, on 27 September 1772, James Brindley died at Turnhurst Hall after a short illness; overcome by diabetes and the pressures of canal construction. Professionally, Hugh Henshall was Brindley's chosen heir. In Brindley's obituary, published by *Aris's Birmingham Gazette*, it is said: 'Mr Brindley had long been sensible of the precarious situation of his Health and wishing to be succeeded in his Profession by his brother-in-law, Mr Henshall, Clerk of Works of the Grand Trunk Navigation, he spared no pains to qualify him for that important Trust.' Henshall also took over the support of Brindley's widow and her two young daughters. Josiah Wedgwood told Sir Roger Newdigate that Henshall's 'knowledge of mechanics, unwearied industry, and strictist integrity' rendered him well qualified to complete the canal, which was now open from 'the Trent to the Pottery'.

The Chesterfield Canal was surveyed by James Brindley in 1769. The line ran 46 miles from Chesterfield to the River Trent at Stockwith, via sixty-five locks and a 2,850yd-long summit tunnel at Hartshill, or Norwood as it now known. The canal was built to the same narrow dimensions as the southern section of the Trent and Mersey Canal until it reached Stockwith in the east, where larger locks were built. Trade focused on the carriage of iron, coal, timber and lead. After Brindley's death, Hugh Henshall took over as supervising engineer to the canal company on 25 November 1772. By 13 May he had agreed to make a complete inspection of the canal every three months and report to company meetings in person. On 2 March 1774, Henshall was appointed resident engineer as well as consulting engineer, as John Varley, the incumbent engineer, was finding supervision of the project difficult and his brothers and father had been found guilty of overcharging the canal company for their contract work in the tunnel. The brothers and father were dismissed but John Varley stayed on to work for Henshall. Hugh reorganised the building operation completely. He had bricks manufactured and marshalled where they were needed, with no surplus, and kept a close check for the company on building and manufacturing expenditure.

By 1773 the full burden of completing Brindley's work, looking after his widowed sister Anne and her two young children, plus his elderly, infirm parents and grandfather, was adding to Hugh Henshall's responsibilities. Now in late middle age, Hugh was also guardian to James Brindley's daughters, Anne and Susannah, was responsible for their tuition and acted as executor to James Brindley's estate. This responsibility was further complicated by the fact that the great engineer had not left a will. Henshall had received power of attorney over Anne's affairs immediately after Brindley's death, and he signed documents for her in this capacity; Letters of Administration having been granted on 18 December 1772 by the Birmingham Probate Registry. There were debts to be collected, and Henshall represented his sister's interests in trying to recover monies owed to James Brindley by Thomas Gilbert for work undertaken between 1762 and 1765, plus monies owed by the Duke of Bridgewater for the work undertaken by Brindley on his canal.

In the same year Henshall was also engaged in correspondence between Thomas Harrison, the agent for Lord Paget, over monies owed for the purchase of land and 'contingent damages' incurred in the construction of the Trent and Mersey Canal. On 10 May 1773 Henshall wrote from Turnhurst Hall to Lord Paget regarding the sums of

Saltersford Tunnel on the Trent and Mersey.

money owed. This indicates clearly that he had taken up residence, even if temporarily, with his widowed sister and family in order to give her support and advice. A year later Hugh's father John Henshall died, and the responsibilities on Hugh increased further. In September 1774 Henshall, still pounding the rutted and muddy roads that Arthur Young had alluded to as 'infernal' and 'impossible to describe in terms adequate to their deserts', was involved in completing the difficult section of engineering between Acton and Middlewich in Cheshire, on the Trent and Mersey Canal. As Josiah Wedgwood wrote: 'Mr Henshall deeming it impracticable to make the canal along the high sloping banks on the side of the Weaver has cut those new tunnels [Saltersford and Barnton].'

In 1775 *Aris's Birmingham Gazette* reported that Josiah Clowes, working for Henshall as a contractor, was advertising for labourers on the Trent and Mersey Canal at Middlewich. This was in order to get the northern end of the canal completed. Middlewich became the transhipment point for barges, allowing for broader and heavier barges belonging to the Duke of Bridgewater to navigate the northern section of the canal. This section had therefore been redesigned with greater dimensions to allow for the duke's craft. On 7 March 1775 Hugh Henshall gave evidence to a parliamentary committee, requesting permission to raise more money. Henshall's response to Parliament indicated 75 miles of the canal were complete and that 'the great Tunnel through Harecastle Hill of the length of about 2,880 yards' was made.

Years passed. Early in February 1776 Hugh Henshall's mother died and was buried at Newchapel on 28 February. On 5 September 1776 Henshall was criticised by the Chesterfield Canal Company for not attending the works as agreed, but in this year the canal was virtually completed. It opened fully for trade on 4 June 1777. In the same year the Trent and Mersey Canal was completed. Henshall had, despite many engineering and personal difficulties, eventually brought James Brindley's major and enduring work to fruition.

Stables, Greenway Bank.

In 1778, Hugh Henshall purchased the farm and adjoining acreage at Greenway Bank – whether he lived there himself is not known. The farm, situated near Knypersley Reservoir and built in the early eighteenth century by Thomas Telford, was in the parish of Norton-in-the-Moors, 3 miles north of Stoke, and had extensive stabling and paddocks. Its purpose was very probably to raise and stable horses for boat haulage and the carrying trade on the Trent and Mersey Canal. (Remains of the farm and stables exist today and are part of a country park.)

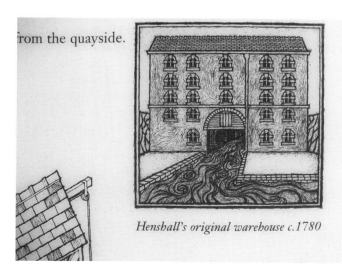

Henshall's original warehouse c.1780

Hugh Henshall's warehouse, Manchester.

Henshall's energies now turned primarily to his business interests, although the canal carrying business had been developed and initially run during his busy years as an engineer. On 23 May 1772 John Wedgwood (probably Josiah's uncle) recorded in his account book that five crates were to go to Stone by 'Mr Henshaw's [sic] boats down the canal'. They were ordered to 'William Fletcher, Warfinger, Gainsborough and to his direction at Hull'. This canal carrying business, 'Hugh Henshall and Co.', became the Trent and Mersey Canal proprietors' carrying company. During 1775 Josiah Clowes rented boats from Henshall at Middlewich on behalf of the Chester Canal Company. Clowes was also working as a canal trader at this time at Middlewich on behalf of the Trent and Mersey Canal Company. The canal trade between south Lancashire and the Midlands became extremely profitable, despite the reputation of some boatmen as 'being none other than inland pirates' and later, links were made with London on the completion of the Coventry and Oxford canals.

Tensions existed, however, between Henshall and Co. and the carrying company of the Duke of Bridgewater, represented by John Gilbert. These were eventually resolved after some acrimony regarding tolls, and John Gilbert shared a warehouse with Hugh Henshall's company in Castlefield, later known as the Grocer's Warehouse (demolished after the Second World War). The site of the old warehouse can be visited in Manchester today.

Henshall's company was eventually taken over by Pickfords, almost certainly on a regional basis. In 1786 Pickfords boats were commuting between Manchester, down the Bridgewater and Trent and Mersey canals, to Shardlow. Goods were then taken by road to London. During the next fifteen years Pickfords progressively adopted canal conveyance between Manchester and London, exploiting the new canals as they became available. Therefore Henshall's early waterway enterprise provides a direct link with modern road carriage.

On 20 September 1790, Hugh Henshall was asked to survey and prepare estimates for the building of the Manchester, Bolton and Bury Canal. This canal linked the River Irwell in Manchester with the growing cotton town of Bolton. The canal was 11 miles long with a 4¾-mile branch to Bury. By 13 October Henshall had produced his survey and the report. In July, the following year, Matthew Fletcher and Hugh Henshall were organising contracts for cutting, and on 10 September 1791, Hugh Henshall was acting as chief engineer. The canal, which was initially planned as a narrow canal, was eventually engineered with seventeen broad locks, 68ft by 14ft 2in. The hope was to connect at a future date with the Leeds and Liverpool Canal. The Manchester, Bolton and Bury Canal was opened completely in 1808.

In 1783 a new wagonway from the Gilberts' Caldon Low limestone quarries was built by Hugh Henshall, linking the quarries to the Caldon Canal. He probably took advice from Thomas Dadford on this arrangement. Cast iron bars were pinned down upon rails of wood fixed across wooden sleepers. This route was improved upon by John Rennie in 1803 and eventually became a railway in 1849 when the North Staffordshire Railway bought the canal.

Hugh Henshall's business developments were complemented by a further move into the business that the canal had been designed to stimulate, namely, pottery. Hugh went into business with Robert Williamson, Anne Brindley's second husband, in a pottery works at Longport in the 1790s. In 1793 the firm was entitled Henshall, Williamson and Clowes, and specialised in the production of black basalt ware. Despite these developing entrepreneurial interests, Henshall was still available as a consultant on canal schemes

Prestolee Aqueduct on the Manchester, Bolton and Bury Canal.

and surveys. In the autumn of 1792 Henshall, now living in Longport, re-surveyed Josiah Clowes' plans for the Herefordshire and Gloucestershire Canal; reporting to the committee on 3 September and 9 October 1792. He suggested a shorter route with the Newent branch incorporated into the main line of the canal. On 6 June 1785 he also gave advice to the Mersey and Irwell Navigation on damage caused by flooding. His report was received by the canal company in December 1787, although business work in Staffordshire meant that similar requests were not always followed up

For the next five years Henshall's efforts were directed to his various businesses and canal carrying enterprise. In 1793 Henshall was involved with canal engineering further afield. On this occasion, he worked with William Jessop on a survey of the Grand Western Canal. The Grand Western Canal was to have been part of an ambitious scheme to link the English and Bristol Channels, thereby obviating the dangerous voyage around Land's End. William Jessop was asked by the company to comment on the two surveys. Interestingly, he appointed Hugh Henshall to the task. Jessop's report to the committee on 28 November 1793, based on Henshall's work, recommended Longbotham's line but was unenthusiastic about the Tiverton and Cullompton branches. The significance of the choice of Henshall as surveyor should not be overlooked. Jessop was clearly one of the outstanding civil engineers in the country at this time and he recognised, by his choice, Henshall as one of the country's most skilled and experienced surveyors of waterways.

In 1794, between April and July, Henshall was assisting the son of another member of Brindley's school of engineering, Thomas Dadford Jnr, in a survey of a short tram road to serve the Brecknock and Abergavenny Canal. The tram road, 1⅜ miles long, ran from Llam March 'Coal and Mine Works' to the Clydach ironworks. It opened in June 1795. Henshall also gave advice to Thomas Dadford Jnr regarding the relocation of the Ashford Tunnel to prevent the possibility of slippage. In this Henshall would have been well qualified, with his experience of engineering the northern section of the Trent and Mersey Canal. Henshall held shares in most of the canal companies that he worked for and the Brecknock and Abergavenny Company was no exception; he held seven shares in 1796.

On the first day of January 1795, it is likely Henshall attended the funeral of his old friend, canal engineering colleague and brother-in-law, Josiah Clowes. The service took place at Norton-in-the-Moors church. In 1799 the Henshall family suffered another unexpected bereavement when Robert Williamson died suddenly, on 3 October. Hugh Henshall had to assume the practical role of head of the family once again, only this time the family businesses were very extensive. In 1806, when Henshall was seventy, remarkably he was still involved in canal work despite his business interests. The Glamorganshire Canal Company made an award to Hugh Henshall on 24 May, with reference to a dispute over water supplies. Henshall had travelled to Wales to adjudicate over this matter. Also, the Manchester, Bury and Bolton Canal remained incomplete. The work was still there and his expertise was still in demand.

Henshall's business interests left him a very wealthy man. On his death in 1816 he left a large number of properties, business premises and shares. These included extensive properties in Wolstanton and Greenway Bank, and mines at Ranscliff and Golden Hill. Businesses listed in his will included pot works, shops and land in Longport, Dale Hall and Burslem. Also included were farms at Biddulph and limeworks at Newbold Astbury, south of Congleton in Cheshire. Further properties belonging to Hugh Henshall included warehouses and premises associated with the Trent and Mersey Canal, plus the Hendra Company, once owned by the Gilberts, which bought clay chiefly from Cornwall.

After Hugh's death, the proceeds of these assets were disposed to fourteen members of the Brindley, Henshall and Williamson families. Special consideration was given to James Brindley's elder unmarried daughter, Miss Anne Brindley, who inherited Greenway Bank as long as she did not marry. In that event the property was to pass to Hugh

Ashford Tunnel on the Brecon and Abergavenny Canal.

Little Ramsdell Hall, home of Anne Williamson.

Henshall Williamson and his heirs. However, the bulk of Hugh Henshall's assets were left to his sister, Anne Williamson, and her sons, Hugh Henshall Williamson and Robert Williamson.

Before his death, on 5 October 1810, Henshall's will was amended to leave more of the assets to Robert Williamson and less to Hugh Henshall Williamson. A second codicil, dated 22 February 1814, referred to properties purchased from John Gilbert. The latter had died in 1812. These properties included the limeworks at Astbury and freehold lands, part of the Gillow estate in Biddulph. The second codicil also referred to the purchase and deeds of the seventeen freehold houses and shops which Hugh had purchased with his sister Anne. The two codicils taken together indicate that Henshall and his sister Anne were trying to secure financial independence for Robert Williamson and Hugh Henshall Williamson for life, and at the same time trying to balance their respective benefits. Undoubtedly, the business acumen of Hugh Henshall anticipated the successes of the Victorian age that was to come.

Certainly the Henshall and Williamson families did very well financially and socially from the canal, trading and pottery businesses. Anne Williamson's son Robert eventually purchased Little Ramsdell Hall in Cheshire. This striking Georgian building overlooked the Macclesfield Canal and was close to the limeworks at Astbury. Hugh Henshall Williamson married Anne Clowes, the great-niece of Josiah Clowes. The couple lived at Porthill in Norton-in-the-Moors and Hugh eventually became High Sheriff of Staffordshire in 1834. Little Ramsdell Hall is now in private ownership but a good view of this imposing property may be gained from the Macclesfield Canal. It is tempting to wonder whether the lost portrait of Anne Brindley/Williamson, *née* Henshall, might have been left there: possibly even a portrait of Hugh Henshall himself.

Grave of Henshall's half-niece, Anne Williamson, owner of Little Ramsdell Hall.

Graves of Hugh Henshall and his sister Anne, James Brindley's wife, St James church, Newchapel.

Henshall family graves, including that of Hugh Henshall, St James, Newchapel. Brindley lies next to them in the distance.

Hugh Henshall was perhaps not devoted to the civil engineering of canals in the same single-minded manner as his two brothers-in-law, James Brindley and Josiah Clowes, or, for that matter, Robert Whitworth. However following James Brindley's death, Hugh Henshall was to spend a further five years engineering the most challenging sections of the Trent and Mersey Canal, namely Harecastle Tunnel, the northern section of the canal and including three further tunnels; thus completing the Grand Trunk Canal as set out by James Brindley and himself. Hugh Henshall was also to survey seven other canals successfully and engineer two further canals. Ultimately, throughout his long life he never abandoned the professions of surveyor and engineer despite his competing business interests.

Furthermore, Henshall developed the carrying trade on the Trent and Mersey Canal and other navigations as well as his pottery business at Longport. He deserves more than his present footnote in industrial history as he had the clear technical ability to bring others' engineering ideas to fruition, to draw their maps and to improve their surveys. He also possessed excellent engineering skills himself, as well as the ability to organise great schemes. It should also be remembered that he was still undertaking canal surveys and associated work well into his seventies. Indeed, Henshall's work spanned the whole era of canal building in Britain. Also of note, it is due to his carrying company that the present road haulage firm of Pickfords has developed. In a sense he capitalised on the new transport revolution of eighteenth-century England and was an early beneficiary of the burgeoning industries and trade that the Canal Age brought to the country. Finally, it is in part due to the efforts of Hugh Henshall that we have the three-dimensional portrait of James Brindley recorded by Samuel Smiles in the *Lives of the Engineers*.

On 16 November 1816 Hugh Henshall died at Longport, having reached the remarkable age of eighty-two. He was buried at St James, Newchapel, near James Brindley and in the same tomb as his second brother-in-law Robert Williamson, and his nephew John Henshall Williamson (who had drowned tragically in an accident in June 1810). His sister, Anne Williamson, was later buried in this same tomb with her brother, son and second husband, on 26 September 1826. She was aged seventy-nine.

There is no known portrait of Henshall, although Smiles refers to a portrait of Anne Henshall. He describes her as 'a comely girl'. (This portrait is untraceable at present.) Clearly, Henshall was an intelligent, hard-working man of high probity, as Josiah Wedgwood had opined. He was also blessed with his family's strong constitution and business sense. His character appears to have been generous and avuncular, given his devotion to supporting his family, sisters, parents and nephews and nieces throughout his lifetime. This may have been one reason why he did not embrace the mantle of civil engineer entirely after the construction of the Trent and Mersey Canal, so that he could be near them. From the interview with Thomas Bentley it might be deduced, given the criticism of Brindley's obsessive working habits, that Henshall believed in the 'agreeable reliefs that are administered by miscellaneous reading, and a taste in the polite and elegant arts'. That is a guess; we do not know for sure. Intelligence, hard work, organisational ability and excellent business sense mark his achievements.

Although Henshall choose the title Esquire for his tomb, and his later interests were predominantly in business and trade, the vital part that Henshall played within Brindley's school of engineering should be fully understood and appreciated. Essentially, he was part of a school that established the foundations of Britain's canal network and its carrying trade.

3

Saml Simcock

SAMUEL SIMCOCK *c.*1727–1804

'His long canals … winding in lucid lines'
– Erasmus Darwin, 1772

Samuel Simcock, foreman carpenter, millwright, surveyor, contractor, engineer and canal carrier, was thought to have been born in Leek. The date of his birth can be calculated from his age at burial, but it was possibly 1727, if not a year later. This places him firmly as a contemporary of James Brindley, who was born in 1716. Both men were millwrights working in the same area and knew one another. Nothing is known of Samuel's parentage at present but he may have had a brother, Johnathan, who also worked as a millwright.

Association with Brindley led Samuel to also make the acquaintance of Brindley's younger sister, Esther. Esther Brindley, or Brundley as she is sometimes referred to in documents, was born in 1718; she was nine years older than Samuel. In 1747, as the result of this relationship, a daughter, Susannah, was born and named after Esther and James Brindley's mother. Susannah Simcock was baptised on 20 February 1748. Sixteen months later, on 12 June 1749, Samuel married Esther Brindley at St Peter's church, Prestbury, in Cheshire. It is most probable that during the period between the birth of her daughter and her marriage to Samuel, Esther lived with her parents. It is equally likely Samuel and James Brindley got to know one another outside their professional lives. What, if any, work they did together at this time is not known, but later, in 1755, we hear of James employing Samuel in mill construction work. Between 1 December 1755 and March 1757, Brindley had been employed in the construction of Ashbourne Mill. In this work we find reference to at least eight men: Thomas Pearson, Thomas Bull, John Thorneycroft, Sanders, John Binat, Joseph Wagstaff, Johnathan and Samuel Simcock. In 1755 James Brindley entered into his Day Book that for one and a half days' carpentering Samuel Simcock had been paid 1*s* 9*d* – a rate of 1*s* per day.

Cyril Boucher records that Samuel Simcock became Brindley's right-hand man, being first of all a foreman carpenter and then a surveyor on the first Bridgewater Canal. As a millwright, like Brindley, he would have become the master of taking accurate levels, directing water, draining, constructing earthworks and diverting flow through dams and sluices. He would also have had the ability to work in stone, brick, iron and wood; all vital attributes for a canal engineer. It is highly probable that Simcock was involved in the work of building the great embankments at Bollin and Stretton on the second Bridgewater Canal. Brindley's methods for building embankments involved using a caisson of planks to

St Peter's church,
Prestbury, where Samuel
Simcock married Esther
Brindley.

Entry from James
Brindley's Day Book
regarding Samuel
Simcock. (Kind permission
by ICE)

THE SIMCOCK FAMILY

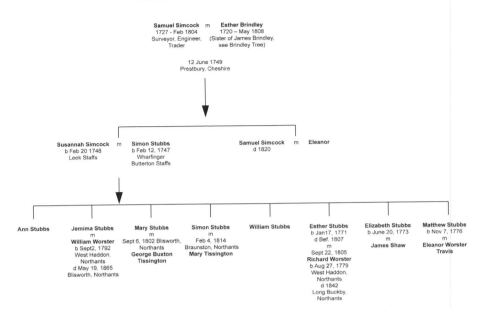

Samuel Simcock m **Esther Brindley**
1727 - Feb 1804   1720 – May 1808
Surveyor, Engineer,   (Sister of James Brindley,
Trader   see Brindley Tree)

12 June 1749
Prestbury, Cheshire

Susannah Simcock m **Simon Stubbs**   **Samuel Simcock** m **Eleanor**
b Feb 20 1748   b Feb 12, 1747   d 1820
Leek Staffs   Wharfinger
                        Butterton Staffs

Ann Stubbs   **Jemima Stubbs**   **Mary Stubbs**   **Simon Stubbs**   **William Stubbs**   **Esther Stubbs**   **Elizabeth Stubbs**   **Matthew Stubbs**
                 m   m   m   b Jan17, 1771   b June 20, 1773   b Nov 7, 1776
                 William Worster Sept 6, 1802 Blisworth, Feb 4, 1814   d Bef. 1807   m   m
                 b Sept2, 1792   Northants   Braunston, Northants   m   **James Shaw**   **Eleanor Worster**
                 West Haddon,   **George Buxton**   **Mary Tissington**   Sept 22, 1805   **Travis**
                 Northants   **Tissington**   **Richard Worster**
                 d May 19, 1865   b Aug 27, 1779
                 Blisworth, Northants   West Haddon,
                                                                                                         Northants
                                                                                                         d 1842
                                                                                                         Long Buckby,
                                                                                                         Northants

An early aqueduct through embankment
on the Bridgewater Canal, River Bollin.

start the canal channel before spoil from elsewhere was brought to build up the structure. This is interesting as it indicates that of all the students in the school, Simcock was the longest-serving member.

During the building of the second Bridgewater Canal, John Phillips reports that the 'Smith's forges, the carpenters and mason's workshops, were covered barges, which floated on the canal, and followed the work from place to place; by which means there was very little or no hindrance of business from accidents.' Whilst Simcock worked on these boats did Brindley, Wedgwood and the Duke of Bridgewater smoke pipes together in the old timbered hall in Worsley, discussing future canal strategy? Samuel Simcock is next mentioned in the will of James Brindley's father, also named James, who on 19 November 1763 instructed his executors that £30 was to be left to his daughter Esther, 'wife of Samuel Sincock [sic]'. James Brindley's father died in 1770 but probate was not granted until after the death of his famous eldest son in 1772. At some stage – the exact date is not known at present – Samuel and Esther had a son also named Samuel.

Simcock took part in the initial survey and trial borings for clay suitable for brick making, or for puddle, for the Trent and Mersey/Grand Trunk Canal early in 1760. Once the work on the Great Cross had commenced, however, Simcock was detailed by Brindley to work on the Staffordshire and Worcestershire Canal. After Hugh Henshall had mapped Brindley's survey in more detail in 1766, Simcock was given responsibility for the setting-out of the canal, working with Thomas Dadford and James Brindley on the construction. The brief story of the work on this canal is to be found on the chapter on Thomas Dadford.

Peter Cross-Rudkin, in his article on the construction of the Staffordshire and Worcestershire Canal, reveals that setting-out the summit level was Simcock's responsibility. However Simcock was not to stay there long, as Brindley, overburdened with the canal system he was planning and building at great speed, was now suffering the acute effects of diabetes. Brindley needed Simcock's reliable knowledge on site to assist with the construction of the Birmingham Canal, a challenging engineering project.

Mason's mark – culvert in Bollin Embankment.

During Simcock's period of time working on the Staffordshire and Worcestershire Canal his daughter Susannah had married Simon Stubbs, who had assisted with the Staffordshire and Worcestershire canal after Simcock left.

On 28 January 1767 the committee of the Birmingham Canal Company met at the Swan Inn in Birmingham and agreed to ask Brindley to prepare a plan and an estimate for construction of a canal crossing the counties of Staffordshire, Salop (Shropshire) and Warwickshire. Brindley produced two plans for the canal which he presented to the committee on 4 June the same year. The plan, eventually adopted, was for a higher route from New Hall over Birmingham Heath, through Smethwick, Oldbury, Tipton Green and Bilston; and from there to the junction with the Staffordshire and Worcestershire Canal at Aldersley. There were two main branches to Wednesbury and Ocker Hill.

After Manchester, Birmingham was becoming the prime focus for the industrialisation of the country, as many of the tools and products to further the commercial development of the nation's industry were manufactured in what was to become the country's second city. Not without good reason did it become known as the City of a Thousand Trades. The General Assembly Minute Book of the Birmingham Canal Navigations, of 9 November 1838, put it succinctly when it recorded: 'The Birmingham Canal, with its immense Local Trade, with its numerous Branches traversing in every direction the richest and most enterprising Mineral District in the Kingdom, is without a parallel, and must be judged of solely, with reference to its own peculiar circumstances.' The city was growing rapidly. By 1800 it was the sixth largest city in order of population in Britain, with 73,670 inhabitants. Other significant towns to benefit from the building of the canal were Wolverhampton, which was thirty-third largest with 12,565 people, and Dudley, forty-ninth with 10,107.

Birmingham was a difficult destination to get to by water, being on the top of a 491ft-high ridged plateau with steep slopes in all directions. However, this was also the reason why a canal would be an effective way of moving heavy goods to and from the town, and to the neighbouring Black Country. Once an Act of Parliament for a canal had been passed, work started immediately. Henshall and Whitworth were drafted to the area by Brindley to re-survey the route and check levels. Clay was sought for bricks, while there was also a call for timber and barrows. Advertisements were published for workmen and Brindley was placed in charge of further operations as surveyor on £200 per annum.

The Birmingham Canal left the Staffordshire and Worcestershire Canal at Aldersley Junction and climbed 151ft through twenty locks to Wolverhampton – later there were to be twenty-one locks. Initially the plan was to tunnel at Smethwick on the summit, but having bored and sunk several pits running sand was discovered, Brindley then gave his opinion that 'the best way was to avoid tunnelling; and to carry the Canal over the Hills by Locks and fire Engines'. Simcock, now acting as resident engineer, was ordered to survey the summit to find further sources of water. He used notched weirs for this work, which he and Brindley would have had experience in using whilst constructing watermills together in the North Midlands.

The Birmingham Canal required heavy engineering. The next level after the twenty locks at the western end was 18¼ miles long. There was then a six-lock rise to the summit. The summit at Smethwick was 1 mile in length, with a 1,000ft cutting 28ft deep. From the summit the completed canal descended 18ft by six locks and 4⅜ miles to Birmingham.

Wolverhampton flight on Birmingham Canal.

After Brindley's initial meeting with the committee on 8 June 1768, concerning the revision to the summit level, Simcock reported his findings to the committee on 22 July. There were still issues with regard to water supply and the estimated number of boats passing the summit. Simcock's report was sent to Brindley to ask his opinion as to how many boats would pass the summit, so that the issue of lockage and water supply could be addressed during the construction. Further difficulties ensued when the line was re-routed from the middle of the Spon Lane Locks to Bilston. Differences in opinion between James Brindley and one of the proprietors of the canal and local mine owner William Bentley, resulted in a 'a very sore quarrel'. By now Simcock's new route had also grown from 6 to 12 miles, presumably to find water, serve mines and increase mileage receipts. The parliamentary plan for the canal was drawn by Robert Whitworth and shows a length of 17 miles, although by the time the industrious Simcock had completed the canal it was fractionally under 23 miles, with a 4-mile branch to the coalfields at Wednesbury and a ⅝-mile branch to Ocker Hill.

S.R. Broadbridge, in his history of the Birmingham Canal, explains that the canal was completed in three stages. The first from the Wednesbury coalfields to Birmingham; the second from outside the city to two termini in Birmingham, Brickkiln Place and New Hall – this following considerable arguments and litigation as to its final destination. The third section was constructed from Spon Lane to Aldersley Junction. When coal started to be transported along the partially completed line from Wednesbury, it was taken to Paradise Street outside the city; consideration still being given to the most suitable terminus. Simcock proposed Summer Row but the committee had purchased land at Brickkiln Place (now Gas Street Basin), whilst the original survey had named New Hall. Simcock designed wharves for the terminus at Brickkiln Place. In the event, after further litigious arguments, the canal used Gas Street Basin and New Hall. To the west, the canal owners also had disagreements as to where it should make a junction with the Staffordshire and Worcestershire Canal, and the latter canal company took out a mandamus, a Superior Courts Writ, to speed up the connection. This had the necessary effect and the canal was complete for navigation by 25 March 1772.

Gas Street Basin, Birmingham Canal. The James Brindley public house in the background is due for redevelopment.

The meanders in the Birmingham Canal were so great that James Brindley and Simcock were required to attend a company meeting in January 1771, here they were asked to sign a declaration that they had never received any instructions to make the canal indirect, and had never done this intentionally, but that the course had been chosen by them entirely in their capacity as engineers. However, they also defended the tortuous route by saying that it served more businesses and thereby generated greater revenue for the canal company. The route from Oldbury to the junction led to the suggestion by a local wit that 'Samuel Simcock had used it as a means of immortalising his initials.'

By this date, Brindley was undoubtedly getting irritated by events relating to the canal, the politics of its construction and its argumentative owners, and he left Simcock to follow his contours without perhaps designing and arguing for a different and more effective engineering strategy to lower the summit level. This strategy was provided at a later date by John Smeaton, with a lower line that conserved water and required less effort, on the part of boatmen, to work. Whatever Brindley's and Simcock's justifications, the canal could and should have been more efficiently engineered and surveyed. Four and half years after construction had begun the canal was opened, on 21 September 1772. 'The Company resolved to pay Mr Brindley to the 29th day of September,' and stated 'that they shall always be desirous of his Advice upon Company's affairs when they may require the Assistance of an Engineer.' This was not to be forthcoming, as James Brindley was to die at Turnhurst Hall six days later. Final and dramatic improvements to the canal were left for Thomas Telford to engineer in 1825.

Simcock completed the Birmingham Canal whilst he was also working, at Brindley's request, as assistant engineer to the Oxford Canal; helping to lay out the line. The Oxford Canal formed the southerly arm of the Great Cross, which linked the Coventry Canal at Hawkesbury Junction and the River Thames at Oxford, thereby providing a through-route to London and the south. Sir Roger Newdigate of Arbury Hall, near Nuneaton, was the leading promoter. Sir Roger was MP for Oxford and owned coal mines in Coventry.

In the summer of 1768 Brindley was asked to survey the route and produce plans for parliamentary approval. Following royal assent, James Brindley met with the canal company's general assembly. At their first meeting, on 12 May 1769 at the Three Tuns public house in Banbury, James Brindley was appointed engineer and general surveyor at £200 per annum. Interestingly, Brindley purchased £2,000 of shares in the canal company, which gives some idea of his wealth accrued over ten years of canal engineering work. James King became clerk of works and Samuel Simcock was appointed as assistant engineer. In typical Brindley tradition, bricklayers and carpenters were sent to canal works in Staffordshire for their improvement, in what Christine Richardson has described as a 'workcamp'. It is possible that Simcock accompanied some of them to supervise their training. In 1769, 700 men started cutting the canal from the Coventry end. The canal was 16ft wide at the bottom and 5ft deep. Hugh Compton, in his history of the Oxford Canal, reminds us that a crucial decision was made at the George Inn at Lichfield on 14 December 1769, when it was agreed that all canals being made in the Midlands would standardise their dimensions. 'It was agreed that the locks would pass boats 74ft long x 7ft wide with a draft of 4ft, 4in.' The Oxford Canal was therefore now part of this standard system.

The Oxford Canal rose 118ft by eighteen locks in 27¾ miles from Oxford to Banbury. From Banbury to the summit at Claydon there was a further rise of 77⅓ft by twelve locks. The summit was 10¾ miles and the canal then fell 55¼ft to Napton by nine locks. A further 16¾ miles took the canal to Hillmorton and then, after ½ mile and a fall of 19ft by three locks, occurred the 26½-mile level to Coventry.

After his death on 27 September 1772, Brindley was replaced by Samuel Simcock. Simcock was now engineer and surveyor to the Oxford Canal on £200 per annum, the same salary as Brindley. He was also now faced with considerable work: making the junction with the Coventry Canal to the north; the 1,138yd-long tunnel at Fenny Compton; a 125yd tunnel at Newbold; and continuing to meet demands for a reliable water supply to feed the summit. On 4 April 1774 the general assembly said they felt that Simcock should also consult with Thomas Yeoman and John Smeaton on the issue of water supply. Yeoman advised Simcock and the assembly that the Byfield Water should be used as a feeder and that the 11 miles of the summit should be deepened by half a yard to provide sufficient water. These recommendations were incorporated by Simcock into the building of the new canal. A reservoir was constructed at Clattercote in Oxfordshire, north of Banbury, which was later to be supplemented by the much larger Wormleighton Reservoir. Robert Whitworth travelled south from Yorkshire to give Simcock advice on the building of this reservoir. However, the supply of water to the 10¾ mile summit level was always a problem. In 1787 the Clattercote Reservoir was enlarged. It should have held 3,222 locks of water but its estimated capacity was ⅔ of this, holding 2,200 locks. In 1790, at its opening, the canal was described as 'a dry ditch'.

In the early 1770s, with his career secure, the provision of a good salary plus the wealth accumulated over the past twenty years of successful canal engineering, Samuel and Esther Simcock moved permanently to Oxfordshire and made it their home. Samuel bought lands and an inn called The Crown, near Bicester. The lands approximated to 17½ acres. This included a yard attached to the inn plus stables, outhouses, gardens and meadowland stretching to the edge of the old Bicester priory. Samuel Simcock also owned other lands in Lower Heyford.

Later, Samuel and Esther moved to Ashgrove Farm, Ardley. The farm, mentioned in the *Victoria History of Oxfordshire*, was originally a Saxon settlement. The Simcocks owned the land which was 4 miles north-west of Bicester and was a highly productive, arable and pastoral farm. There the Simcocks had a valuable herd of cattle. The new system of enclosures was benefiting landowning farmers and Simcock was now a considerable landowner, as well an engineer and trader on the Oxford Canal. By the time the farm was passed to the Duke of Marlborough later in the nineteenth century, it occupied 522 acres; the main crops being wheat, barley, oats and turnips.

Simon and Susannah Stubbs had by that time moved nearer to her parents. Simon was benefiting from the growing canal business; he was running a coal wharf and had become a carrier on the Grand Junction Canal at Blisworth. The relationship to his father-in-law's business is not known at present, but it is difficult to believe they were working in isolation. Samuel and Esther's son, Samuel Jnr, had also moved to the area as he was living in Lower Heyford. Whether he was farming, or working in some capacity relating to the canal or the carriage of goods is not known, but canal-related work seems probable.

By 30 March 1778, the first barge loaded with coal reached the wharf at Banbury. The canal company's coffers were depleted at this time. Although limited trading had started at the northern end of the canal, the link between the Coventry–Oxford and Trent and Mersey canals remained incomplete and the real potential of the Oxford Canal lay unrealised. This shortage of funds led the canal company to look at alternative routes to Oxford. There was a suggestion that the River Cherwell be adapted for navigation, but the company sought a different route. In January 1779 Robert Whitworth and Samuel Simcock undertook a new survey. The canal company initially wanted John Gilbert to assist Whitworth – which raises questions about their faith in Simcock at this time, or indicates that Simcock was busy developing trade on the canal – but Gilbert was unavailable and they did in fact use their resident engineer.

In addition to the problems of the landlocked northern section of the Oxford canal, there was also the question of severe economic restraint during the American War of Independence, which led to lack of capital for investment. Indeed, insufficient funds were available to pay off the interest on the loans for construction. Eventually the Duke of Marlborough came to the company's assistance, but it was not until June 1782 that a scheme was agreed. The Oxford Canal would continue its line south and the Trent and Mersey Canal, in conjunction with the Birmingham and Fazeley Canal, agreed to take over the Coventry Canal's rights in order that the link to the north could be completed. In 1783 the American war came to an end and monies became available again for major investment in civil engineering projects.

James Barnes (1739–1819) an engineer born in Banbury who, rather eccentrically, combined brewing with engineering, had now joined the Oxford Canal's team of Whitworth, Simcock and Samuel Weston. Simcock brought the canal through Kidlington on the outskirts of Oxford and he and Samuel Weston produced plans for the final route, to a new wharf near Hythe Bridge in the town. By 1 January 1790 the canal was complete and, to the familiar, welcoming tunes of brass bands, the first barges of goods were unloaded in Oxford. Simcock and Samuel Weston, with four sleeping partners, were the contractors to this extension of the Oxford Canal. Originally they offered to undertake the work for

£29,000 if they were given the monopoly of trade on the canal. The canal company found they had no powers to do this so it became a conventional contract. On the completion of the Oxford Canal the continuing problems of the navigability of the Thames became much more apparent, and Simcock was directed to solve the problem associated with the passage and flooding at Godstow Lock. He completed this work after the Duke of Marlborough advanced further funds to the general assembly.

Early in 1789 the Oxford team of Simcock, Weston and Barnes were called upon to survey the Western Canal, which was the forerunner to the Kennet and Avon Canal. This team was a second-choice group of surveyors and engineers. Robert Whitworth, the preferred surveyor, was unable to attend given his commitments to the Forth and Clyde Canal. The Oxford team proposed several routes which went further north than the canal eventually built, travelling through Calne, Chippenham and Laycock. The plans included the option of a 3-mile tunnel at the summit level, or a higher summit with a shorter tunnel. Nevertheless, the work illustrated the status of these engineers, including the veteran Simcock.

On the completion of the Oxford Canal in 1790, local farmers could sell their produce further afield. This was good news for Samuel Simcock as he had by now obtained a private monopoly on trade on the canal and, with the exception of Cropredy, ownership of the canal company's wharves. Unfortunately, it was bad news for the poor in the area. Hardship resulted due to an absence of cheap local food which was being exported to the Midlands at increased prices. Conversely, the poor did benefit by receiving cheap coal from Wednesbury, brought by Simcock's two canals. New building material in the shape of brick and blue slates, carried by canal, was also making an appearance in the area.

The episode with the lock at Godstow had highlighted the main problem that was evident in the Thames Navigation south of Oxford to Staines. Matters now became of such concern that Simcock and Samuel Weston were authorised by the canal company to survey another possible canal from the Oxford Canal at Thrupp, southwards to Brentford on the Thames, to bypass the unreliable river. The canal was known as the Hampton Gay Canal, or the London North Western Canal, and would have been 61 miles in length, passing the chalk hills by a long tunnel at Wendover. A 3-mile branch to Aylesbury was also part of this plan. This proposed canal was favoured by the Oxford assembly, who were worried by the news that a more direct canal was being proposed to transport goods from the Midlands to London, thereby bypassing their canal. This rival canal, from Braunston to Brentford, was named the Grand Junction Canal (and was eventually completed by William Jessop and James Barnes in 1805).

On 11 February 1793 a bill for the Hampton Gay Canal was presented to Parliament. Samuel Weston had undertaken the survey and his plans and arguments were supported in Parliament by Samuel Simcock and Hugh Henshall: a meeting of the old school of engineering took place. Doubtless there was much conversation about the carrying trade and the economy, as well as engineering issues. Despite the money that Simcock, Henshall and some of the other members of Brindley's school had made, the value of money had fallen by 40 per cent in the years between 1780 and 1803. The uncertain situation in Europe after the French Revolution and the Napoleonic Wars, not yet over, had caused severe inflation in England. Whether, for instance, Simcock and Henshall could afford a visit to the theatre like their old mentor and joint brother-in-law is not known. London productions were not only frowned upon by James Brindley; R.R. Angerstein

Cottage on the Oxford Canal at Lower Hertford; note the brick and slate construction.

thought that the *Beggars Opera* at Covent Garden Theatre served only 'to encourage highwaymen and robbers'.

This is the last time that we hear of Samuel Simcock with regard to his canal work. Although Simcock was a contractor on the canal with Samuel Weston and his son, William, it is likely that in 1794, at the age of sixty-six, he took partial retirement and turned his attention to his canal carrying business and farming (one of the witnesses to Simcock's will, John Chamberlain of Cropredy, was a grazier). In 1794 Simcock had made his will, but he lived for a further ten years at Ashgrove Farm in Ardley, presumably enjoying retirement, before dying in February 1804. He was buried in the churchyard of St Mary's, Ardley on 1 March 1804, aged seventy-six.

In his will, Simcock made considerable provision for his daughter Susannah and her son, Samuel Stubbs. This was to be achieved by annuities to be paid from the rent of his lands in Bicester, and elsewhere in Oxford, to Susannah. His own son, Samuel, was to inherit the lands and farm after his mother's death. Motives can only be guessed at. Did Samuel Snr know that Susannah's husband Simon was ill, and wanted to ensure his daughter's financial independence? Simon died only three years after Samuel himself, and Susannah Stubbs had six children that needed support. Simcock's executors were also instructed to buy shares in the 'Oxford Canal Navigation' and with the proceeds pay an annuity of £50 per annum to Esther.

Esther, in addition to her annuity and Ashgrove Farm at Ardley, was left the livestock, cattle and husbandry tools. These were to be passed to her grandson, Simon Stubbs Jnr, on Esther's death. In 1804 Esther had reached the remarkable age of eighty-six, and was rather unlikely to be actively involved in any strenuous farming activity. However she survived her husband by a further four years. On 20 March 1804, a month after Samuel's death, she made her own will in which her extensive estates in Lower Heyford, Ardley and Bicester were left to her dear son Samuel Simcock. In an interesting preface to the will she describes herself as 'being far advanced in years but of sound and disposing mind, memory and understanding thanks to Almighty God'. Esther was a remarkable woman. She survived until she was ninety years of age, nearly twice the age of her famous brother. She died at Ashgrove Farm in May 1808 and was buried in St Mary's church, Ardley on 7 May 1808. Her grandchildren benefited not only through her and Samuel's bequests, but also through the heritage of well-paid employment as carriers, publicans and coal merchants. This employment was brought into being by the new canal system that Samuel had engineered with his long-dead brother-in-law.

Although little was known until recent researches by Sue Hayton into Simcock's later life, he has perhaps been underestimated in the amount and the importance of his mechanical and civil engineering work during the eighteenth century. Brindley recruited him at an early age in mill and canal construction – on the Bridgewater and the Trent and Mersey canals – and he stayed throughout his working life labouring on the majority of Brindley's projects. Simcock, like Brindley, had moved from mechanical engineering with the construction of watermills, to civil engineering and the survey and construction of canals. Brindley clearly thought highly enough of Simcock's practical abilities to take him to engineer the Bridgewater Canal from the outset, and that cannot simply have been because they were related by marriage. Following the success of that pioneering venture, Simcock moved on to work on every arm of the Great Cross in an important capacity. He was working on the Trent and Mersey Canal from 1760, on the Staffordshire and Worcestershire Canal from 1766 and, finally, on the southern arm of the Great Cross on the Oxford Canal in 1768, training and trained by others all the time. In addition he also engineered the difficult route of the Birmingham Canal from Aldersley Junction to the city, completed in 1772. In all these endeavours he was principally acting as surveyor and resident engineer.

The last two canals he built were tortuous in design and he was roundly criticised for this. However it must be remembered that some of the blame for this lay with the dictates of the canal companies themselves, who avoided expensive engineering costs by building on the contour level, or for political reasons. The latter is exemplified by the Birmingham Canal and the sensitivities of its proprietors with regard to the proximity of the canal to their coalfields or their lands. Although it was left to Brindley, followed by Whitworth and Henshall, to present detailed information and drawings to Parliament and solve the more acute engineering problems, it would seem that Brindley placed complete faith in Simcock's practical ability to manage the majority of affairs at water level. In 1793 Simcock was to go to Parliament to argue the case for the Hampton Gay extension of the Oxford Canal with Hugh Henshall and Samuel Weston, but this was not perhaps his primary strength. He was the man James Brindley trusted to get the job done on the ground.

A fitting obituary for Samuel Simcock can be found in the definition propounded by Rees for the qualifications for becoming a resident engineer:

> … none but the men of strictest integrity and extensive knowledge ought to be employed as resident engineers, and those that the Committee and principal engineer ought not to hesitate in offering and paying such men a very liberal salary to engage the whole of their time; that too great a length of line or extent of business should not be put on such a man.

The unassuming Samuel Simcock played a key role in early canal construction and continued the traditions of the Brindley school of engineering into the last quarter of the eighteenth century. He was rewarded with a wealthy retirement paid for by his engineering skills, his canal carrying trade and his farming activities. He ended his days with Esther on Ashgrove Farm amongst the pastoral beauty of the Oxfordshire Wolds, around which his final canal rambled; a useful life.

4

Robt Whitworth

ROBERT WHITWORTH 1734–99

'Question surveyors, know our own estate'
– Shakespeare, *Henry IV*

Robert Whitworth was arguably the most successful civil engineer belonging to Brindley's school of engineering. He was born in 1734 in Calderdale, Yorkshire. His exact date of birth is unknown but he was baptised on 15 November 1734 in Sowerby Chapel. He is perhaps the one practitioner in this school of engineers worthy of a separate book. By the end of his engineering life, he occupied a place in the minds of canal companies not dissimilar from that of James Brindley himself, and he was as sought after for his advice and expertise. Through his professional life he remained completely devoted to surveying and engineering canals and was responsible for completing two of the great cross-country routes in the north of England and Scotland, as well as assisting with the completion of the Great Cross, other ancillary canals and the improvement of river navigations.

Robert was the eighth of nine children born to Henry Whitworth and Mary Crowther, who both came from Ripponden but were married on 6 June 1716 at Elland, in Yorkshire. Robert's father was born in Ripponden on 20 October 1694 and became a combsmith. The name defines the job but not the specialist nature of the labour; these combs were unique tools used in the textile industry to tease out the individual strands of the exceptionally long wool used for the production of worsted cloth. The combs were made from forged steel. It was a specialist industry in which, if you were successful, the rewards were large, given the demands for strong, good-quality cloth by an ever-increasing population at the start of the eighteenth century. Henry's business made him a wealthy member of middle-class society. In 1722 Henry moved to a house known as Waterside, or Wheatley Royd, in the Blackwater area of Sowerby. The house still stands today and it was to here Robert Whitworth returned after his baptism. Robert's older brother William joined the family business and their father Henry was successful enough in this specialised area of work to purchase other properties to expand his business and, very possibly, to develop sheep farming in the area.

Nothing else is known of Robert's early life. There were six older siblings living at Waterside, four boys and two girls. An older brother, Roger, had died in infancy two years before Robert's birth. In 1736 a younger sister, Mary, was born. It is likely, given Robert Whitworth's considerable mathematical, scientific and literary ability, that he was educated locally in a private school or, in the manner of the time, at a friend's home. A love of learning remained with him throughout his life, as we know from his extensive library and his collection of mathematical and philosophical instruments. Unlike Hugh Henshall whose surveying skill was inherited and nurtured by his father, nothing is known about Whitworth's training until we hear of him in 1761, practising as a qualified land surveyor.

G.W. Oxley, in his excellent transcription and papers on Robert Whitworth, tells us that the earliest example of Robert's work was a plan of an estate in Erringdon. In 1764 Robert was employed by George Stansfield to measure the road from Halifax to Lamhills and Hoyland Mills; this to calculate how much was owed to the contractor during building work at Sowerby church. Robert was paid 7s 6d for this undertaking. However, far more significantly was the record that showed Robert had been measuring and surveying the River Calder and Halifax Brook in 1765, for which he was paid 5s. This was Robert's first known experience of working on waterways, and he is likely to have observed John Smeaton engineering the Calder and Hebble Navigation before James Brindley replaced Smeaton in 1765. Brindley met Robert Whitworth, and this meeting was to create a crucial alliance in the history of Britain's waterway construction. Brindley, fresh from the triumph of the Bridgewater Canal, was the engineer all canal companies wanted to recruit for their new schemes. As Oxley has said, 'Brindley had a good eye for the lie of the land but insufficient literary skills and penmanship for mapping.' This was something that the young surveyor, Robert, certainly possessed. It was also in 1765 that another crucial alliance was forged by Robert, when he married Sarah Irvine from Earsden in Northumberland on 26 December 1765, at Sowerby Chapel.

Robert's painstaking draughtsmanship, mathematical and mapping skills must have been a blessing for the older engineer, as their skills complimented one another completely. In 1766, Brindley and Whitworth were working together on Sir John Ramsden's canal

Wheatley Royd House, Sowerby, the probable birthplace and home of Robert Whitworth.

THE WHITWORTH FAMILY

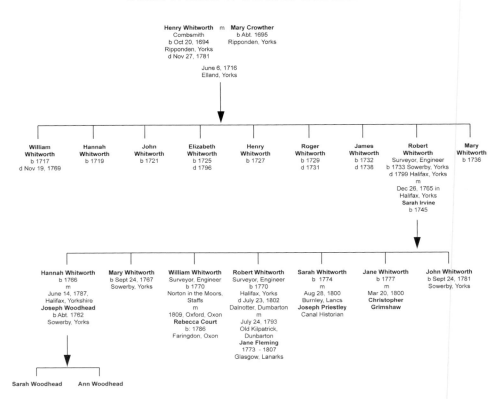

Henry Whitworth m **Mary Crowther**
Combsmith
b Oct 20, 1694
Ripponden, Yorks
d Nov 27, 1781

b Abt. 1695
Ripponden, Yorks

June 6, 1716
Elland, Yorks

William Whitworth b 1717 d Nov 19, 1769	**Hannah Whitworth** b 1719	**John Whitworth** b 1721	**Elizabeth Whitworth** b 1725 d 1796	**Henry Whitworth** b 1727	**Roger Whitworth** b 1729 d 1731	**James Whitworth** b 1732 d 1738	**Robert Whitworth** Surveyor, Engineer b 1733 Sowerby, Yorks d 1799 Halifax, Yorks m Dec 26, 1765 in Halifax, Yorks **Sarah Irvine** b 1745	**Mary Whitworth** b 1736

Hannah Whitworth b 1766 m June 14, 1787, Halifax, Yorkshire **Joseph Woodhead** b Abt. 1762 Sowerby, Yorks	**Mary Whitworth** b Sept 24, 1767 Sowerby, Yorks	**William Whitworth** Surveyor, Engineer b 1770 Norton in the Moors, Staffs m 1809, Oxford, Oxon **Rebecca Court** b: 1786 Faringdon, Oxon	**Robert Whitworth** Surveyor, Engineer b 1770 Halifax, Yorks d July 23, 1802 Dalnotter, Dumbarton m July 24, 1793 Old Kilpatrick, Dunbarton **Jane Fleming** 1773 - 1807 Glasgow, Lanarks	**Sarah Whitworth** b 1774 m Aug 28, 1800 Burnley, Lancs **Joseph Priestley** Canal Historian	**Jane Whitworth** b 1777 m Mar 20, 1800 **Christopher Grimshaw**	**John Whitworth** b Sept 24, 1781 Sowerby, Yorks

Sarah Woodhead	Ann Woodhead

Another view of Wheatley Royd House.

which linked the Calder and Hebble Navigation to Huddersfield. Oxley observes that this canal was attributed to both Brindley and Whitworth. He wonders whether Brindley surveyed the work but left the matters of detail chiefly in the hands of Robert Whitworth. Whitworth was proving rapidly that detail was his forte. Such arrangements seem to form the basis of their working practice for the next six years. In 1766 Robert Whitworth became a father when his daughter Hannah was born, presumably at Waterside House.

Matters now moved on apace and 1767 was a critical year for Robert Whitworth's future career as an engineer and surveyor. Whitworth decided to work full-time with James Brindley. Robert and Sarah then moved to Norton-in-the-Moors in Staffordshire so that he could be close to Brindley and the burgeoning canal world, centring on the vortex of the Great Cross of canals, the Grand Trunk Canal. Norton lay adjacent to Harecastle Hill and the canal workings, as well as Turnhurst Hall where James and Anne Brindley were living with their young family. At Norton, Whitworth must undoubtedly have come into contact with other pupils from the school: Hugh Henshall, Samuel Simcock and Josiah Clowes. Whitworth was later to recommend Josiah Clowes to the Thames and Severn Canal as engineer, and Hugh Henshall occupied Whitworth's role in a more localised form on the Grand Trunk Canal, where he was mapmaker and resident engineer.

Robert Whitworth was now James Brindley's chief surveyor and draughtsman working on the Great Cross and its extensions. He produced survey drawings of the Birmingham Canal. This canal led Birmingham to become one of the most important cities in Britain's rapidly expanding industrial state. Amongst other products, Birmingham made the tools

Statue of Brindley at Coventry Basin.

that were used to create the revolution in production and power across the country. In Birmingham Whitworth worked closely with Samuel Simcock who was engineering the canal. The school was now a force working in reasonably close proximity across the Midlands, with regular forays to London led by James Brindley seeking parliamentary approval for various schemes.

In 1767 Robert Whitworth also produced drawings for James Brindley, of the section of the Coventry Canal which joined the Trent and Mersey Canal as part of the southern arm of the Great Cross to Fradley Junction. That same year Whitworth drew plans of Brindley's survey for the Droitwich Canal, which linked that town to the River Severn; John Priddy was the resident engineer. Again we observe Brindley and Whitworth working together effectively. Brindley was a good spokesman and advocate of the canal idea, as witnessed by his dramatic performances before Parliament. Whitworth on the other hand was 'extremely modest and unassuming … for what Nature had denied him eloquence and colloquial rhetoric, she more than amply remunerated him in the lucid powers of his pen.'

By 1767/8, Robert Whitworth, described as Brindley's clerk, attended parliamentary sessions in order to help with the giving of evidence. He also produced his maps of Brindley's surveys to support arguments about the development of the new central canal system linking the principal rivers in England. By the middle of the eighteenth century London was a city of 700,000 people, but quite primitive. At the time Brindley and Whitworth visited there were only two bridges: London Bridge and Westminster Bridge. Traitors' heads were still displayed as a deterrent at the Temple Bar. There were no pavements on the streets and despite the building that was taking place, the German physicist Georg Christoph Lichtenberg, who visited London in 1770, reported having to write by candlelight at 10 a.m., such was the gloom. Dr Johnson, whose views on London life are well known, took up carrying a cudgel after being attacked by footpads, or muggers.

In 1768 Brindley and Whitworth were involved together again in the survey of a canal from Stockton to Darlington, designed to transport coal to the Tees at Stockton. The same pattern was followed: after a survey by Brindley, Whitworth drew the plans for parliamentary approval. In this case approval was not forthcoming but the route chosen was followed, in part, at a later date by George Stephenson when he built the Stockton–Darlington Railway in 1825. The year 1768 also saw Whitworth surveying and writing a report with an illustrated plan of the intended navigation between Lough Neagh and Belfast. The thirty-eight-page report, presented to the local committee at Hillsborough, gives some idea of the accuracy and mathematical ability that underpinned Whitworth's work. A quotation from Whitworth's report underlines this point:

> Head level to be made 30ft deep of Water, except in steep and difficult places … The mean Breadth of the uppermost foot is Water is 46 feet and half, the length of this level 43,110 feet, so that to sink the Canal one Foot, viz to reduce the Water to 5ft, equal to the Depth of the rest of the Navigation, 2,00,4615 cubic feet of Water may be drawn off, which will supply 331 locks full of Water (supposing the Locks to be made 65ft long, 15 Feet and half broad, and six Feet deep) consequently will discharge half that Number of boats.

The prose may be functional, with a few errant capital letters, but the mathematical calculations are impressive.

In the same year, Robert Whitworth became involved with the first of the major canal projects with which we associate his name; that of the surveying and engineering of the Leeds and Liverpool Canal. The following year Brindley, unable to check the existing surveys of the route, directed Robert Whitworth to examine these surveys, one of which Brindley had been earlier involved in with John Longbotham, a local surveyor and engineer. The Lancashire portion of the committee had requested John Eyes and Richard Melling to survey a more southerly route to the canal their side of the Pennines, so as to incorporate more Lancashire towns. Whitworth discovered considerable flaws in the Eyes/Melling survey. Between Rishton and Burnley, for example, there was a discrepancy of 35ft in planned levels. Although it was possible to overcome the difference, heavy engineering costs would be incurred. The Yorkshire committee rejected outright their Lancashire partners' proposed route, stating loftily that 'no one but Longbotham, Brindley or Whitworth should be employed in survey work'.

This stand-off between the Lancastrians and the Yorkists over the proposed route of their canal finally led them again to ask James Brindley to adjudicate matters. Whitworth might have been well placed, for personal as well as professional reasons, to undertake the work, but Brindley had sent his first-choice surveyor to the West Country for three months to survey a possible canal route to connect the English and Bristol Channels. (This was to improve the time and safety to shipping circumnavigating Land's End.) John Varley undertook the work for Brindley and the canal company instead. Although Whitworth probably had his immediate family based at Norton-in-the-Moors at this juncture, it would have been possible to return to Calderdale to be near the rest of the family had that work been delegated to him.

Sarah Whitworth gave birth to her eldest son, Robert Whitworth Jnr, in 1770. His baptism is recorded at Sowerby Chapel. Some records describe William Whitworth, her second son, as being born in that year. Other records show he was baptised at Norton-in-the-Moors in 1772. This seems more likely and dates from the period of time that Norton was home to the Whitworth family. Both Robert and William became engineers in their father's footsteps. However, neither obtained their father's status and there were issues raised concerning Robert's effectiveness (see the Ashby Canal). This is perhaps why Robert Whitworth Snr took time off his own projects later to help his sons.

Clearly, home for Robert Whitworth was at coaching inns near his canals under survey or construction. He was in Hampshire surveying and mapping the Andover Canal for James Brindley in 1770, before returning to London with Brindley to present evidence to Parliament the following year. It is probable that he stayed at the Kings Arms Inn, Leadenhall Street, at this time, which was his favourite haunt over the next ten years when he attended Parliament, to give evidence on canal construction, or went to Smeatonian Society meetings at their venue at the King's Head in Holborn.

The Great Cross linking the four main rivers of England was only going to be successful if the rivers themselves were navigable. In 1770 Brindley was hard at work undertaking surveys to improve the Thames Navigation. Robert Whitworth produced the report and maps for Parliament of a proposed lateral canal from Isleworth to Monkey Island. The two engineers also surveyed and mapped the line from Monkey Island to Reading. Unfortunately, the schemes were not approved and it wasn't until the Basingstoke Canal, with proposed extensions to Newbury, was built in 1794 that a lateral canal of sorts bypassed this tortuous length of the Thames.

On 29 March 1771 Robert Whitworth's developing engineering abilities were formally recognised when he was admitted into the Smeatonian Society. On that day Whitworth was admitted to the society with John Golborne and James Black; Mr Robert Mylne was chosen as vice president. It was only the society's second meeting held at the Kings Head, Holborn. Between 1771 and 1789 Whitworth attended twenty-two meetings of the society. These professional meetings were to discuss matters relating to the building of canals prior to the giving of evidence to Parliament. On 24 April 1780 Robert Whitworth acted as secretary for the evening. On 12 April 1771 Whitworth also renewed his acquaintance with another Brindley graduate, Hugh Henshall, when the latter was admitted to the society. He also met Thomas Dadford on 25 April 1783 when he too was enrolled as a member of the society. By 6 January 1792 Robert Whitworth was listed as a country member, as by then he was undertaking projects the length and breadth of Britain. Whilst he attended society meetings his address was given as 122 Leadenhall Street.

The year 1771 was also the year that Robert Whitworth was approached by *The Gentleman's Magazine* to provide their readers with information about canals under construction and planning in the country upon which he and James Brindley were engaged. Was Whitworth perhaps becoming the advocate for the 'canal idea', building upon Brindley's advocacy of the previous decade? It was just as well he was on hand to do so because the partnership was about to end. James Brindley died at Turnhurst Hall on the 27 September 1772. Brindley's work was unfinished but, like all great leaders, he had provided a legacy in the school of engineers that he had assembled, trained and tutored. These new civil engineers were in place to finish the task and Robert Whitworth was in the vanguard of these experienced engineers able to meet the canal companies' and country's demands.

Between 1772–74, apart from two breaks in the Midlands and the North, Whitworth was primarily concerned with surveys and engineering work near London. In 1773 he was surveying a canal from Waltham Abbey to Moorfields. Promoted as the 'Grand Commercial Canal', its purpose, never realised, was in part to bypass the lower reaches of the Lee Navigation and included a branch to Marylebone. The canal was a forerunner, in its southern reaches, of the Regents Canal. By 1774 Whitworth had been appointed surveyor to the Thames Navigation, with responsibility for improvements to the river between the City of London and Staines. Originally, he and Thomas Yeoman were appointed to this work together, but Whitworth remained as consultant surveyor to the Navigation for ten years until 1784. He was paid £250 per annum for this work. This was crucial work as the Great Cross of canals would not realise its potential if there were obstructions to trade in the South East of the country and London. In the same year Whitworth undertook two surveys in the North. The first was a survey of the River Trent at Newark. In this work he was assisted by John Grundy, also a member of the Smeatonian Society. The second project involved a return to the North with a survey of a canal to join the Leeds and Liverpool Canal from Kendal. This line eventually became the Lancaster Canal. Finally, he returned to the South to survey a canal planned to link Maldon and Chelmsford.

The distances Whitworth travelled with this work were formidable, and one has to consider the wear and tear on the engineer travelling by horse and perhaps on occasions, by stage coach. The state of the roads, despite turnpikes at this stage of the eighteenth century, was poor. Brindley's Day Book records that it took him seven days to travel from the 'Duke's canal' in Manchester, to London. Most turnpikes were negotiated at 2*d* per mile but then there was the cost of accommodation and food; a good meal costing 8*d*. There were also the dangers of robbers. Dick Turpin had only recently been apprehended and the cities contained footpads.

Economic depression had followed the War of American Independence and insufficient monies were available to finance canal development. Indeed, some canal projects were frozen for the best part of two decades; one such example being the northern section of the Coventry Canal. In 1777 Robert Whitworth reported on the Herefordshire and Gloucestershire Canal. Two years later Whitworth produced a report and survey of a canal from Bishops Stortford to Cambridge, which in the event was not constructed. The year 1779 also saw Whitworth working on the parliamentary plans for the Oxford Canal between Banbury and Oxford. Cyril Boucher comments that 'none of the exaggerated loops were on Whitworth's plans which were 82 miles, 9 miles shorter than the route as made and 5 miles longer than the present reconstructed canal'.

In 1781 a survey was completed under Whitworth's direction of a canal from Ashby-de-la-Zouch to Griff, near Coventry. Later this was to form part of the Ashby Canal, engineered by Whitworth and his older son Robert in 1794; Robert Snr was the consulting engineer. Eventually both men were dismissed as the canal company did not think enough of their engineers' time was being spent on the project. Unsurprisingly, Robert and son were busy elsewhere.

In 1782 and 1783 Robert Whitworth was focusing his attention on cross-country routes in the south of England. The first was a canal to link the Thames and Severn more directly. The route, once surveyed and built, linked the Stroudwater Navigation to the Thames at Lechlade, piercing the Cotswold scarp at Sapperton. The second was to survey a route from Abingdon on the Thames to link with the Kennet and Avon Canal. This canal had a branch that joined with the Thames and Severn Canal east of Sapperton. The story of the Thames and Severn Canal is dealt with in greater detail in the chapter on Josiah Clowes, as Whitworth, in the event, only surveyed the summit level through Sapperton Tunnel. Although Whitworth's name is associated with the laying out of this waterway, Josiah Clowes, recommended by Whitworth, was left to wrestle with the unruly contractors and unrepentant geology of this difficult project. Whitworth referred to the rock on the summit level as 'bad, rocky ground, the very worst I have seen cut through any length'. The other route became the Wiltshire and Berkshire Canal, and was to be surveyed and engineered by Whitworth and his younger son, William.

Robert Whitworth did not engineer the Thames and Severn Canal as is sometimes claimed. He had been offered the opportunity to work on the first of his great engineering undertakings; the completion of the Forth and Clyde Navigation which linked the east and west coasts of Scotland via Glasgow. During 1783 he also found time to survey the Coventry Canal from Atherstone to Fazeley, in order to complete Brindley's planned link from the Trent and Mersey Canal south, via the Coventry Canal and Oxford Canal, to London. His work continued during 1793 for the Thames Navigation and, whilst undertaking duties on their behalf, he managed to get himself arrested as a possible French spy in Gravesend. The activities of a surveyor laid themselves open to suspicion by the authorities, particularly during a time of tension with France where the authorities suspected industrial and political espionage. It should be noted that, by this time, Britain had a tremendous industrial lead over most of Europe. Chrimes notes that a succession of continental engineers visited the British Isles to familiarise themselves with the 'latest developments in the structural use of iron, building construction … and canals'.

A canal linking the Firth of Forth and the River Clyde had first been proposed in the reign of Charles II. After several other suggestions, the Board of Trustees for the Encouragement of Fisheries, Manufactures and Improvements in Scotland, created by the State in 1725, approached John Smeaton to survey such a link. During 1763–64 Smeaton completed this work. Two years later nothing had been achieved so the tobacco merchants of Glasgow on the west coast commissioned Robert Mackell and James Watt to re-examine Smeaton's route so that it might connect to Glasgow. This proposed canal was to be narrower and shallower than that actually built.

Echoes of the disagreements between the York and Lancaster partners of the Leeds and Liverpool Canal come to mind, especially when one discovers the views espoused by the Edinburgh supporters when they learnt of a narrow canal being mooted. Jean Lindsay, in her detailed book on the Canals of Scotland, quotes one contemporary writer as referring to such a scheme as creating a:

> …ditch, a gutter, a mere puddle which would serve the purposes of trade but not the magnificence of national honour … what is commerce to the City of Edinburgh? Edinburgh … is the metropolis of this ancient kingdom, the seat of Law, the rendezvous of politeness, the abode of taste, and the winter quarters of all our nobility who cannot afford to live in London; and for these and other reasons equally cogent Edinburgh ought to have the lead upon all occasions. The fools of the west must wait for the Wise men of the East.

Mackell, whose background was similar to James Brindley, in that he had built water mills and steam engines with James Watt before undertaking canal construction, eventually joined with Smeaton to again revise the route and dimensions for parliamentary approval. After much painstaking survey work, Mackell re-routed the canal on a modified course from Grangemouth to the Clyde at Dalmuir, with a branch to Glasgow. This line was later revised a second time, again taking a more southerly course through Stockingfield, nearer to Glasgow, with a large aqueduct planned to cross the River Kelvin.

In the spirit of the 'Wise men of the East', Smeaton and Mackell eventually engineered a broad canal to Stockingfield, east of Glasgow, but work stopped in January 1775; the canal company owing £31,000. Mackell died in November 1779 and due to the depression in trade that had halted works, the extension to the west remained unfinished. When funds became available again six years later to complete the project, the Forth and Clyde Navigation Company appointed Robert Whitworth, in June 1785, to undertake the work; one branch to the River Clyde and the other to Glasgow. This was a major engineering enterprise, a ship canal 56ft wide and with locks 74ft x 20ft. The engineering necessitated the crossing of the valley of the Kelvin on a massive aqueduct bearing water 8ft deep. It was the largest scale work of this nature to be undertaken to date in the Britain and tells us something of Whitworth's reputation. The typically modest Whitworth referred to the aqueduct as 'new and out of the common road of Bridge building'.

Robert Whitworth immediately proposed changes to the original plans. He recommended the cut at Grangemouth be deepened and widened, and Bowling was

Two views of the Kelvin Aqueduct, on the Forth and Clyde Canal.

chosen as the point of entry to the Clyde. Whitworth also instructed that the canal be deepened to 8ft throughout in order to provide sufficient water. After considering the creation of additional reservoirs in the centre of the route at Dullatur, to ensure the necessary water supplies, Whitworth eventually elected to feed the canal with sufficient water from three reservoirs connected to the Monkland Canal which would act as a feeder at the western end.

Whitworth moved to Scotland where his eldest son Robert joined him on the works. It is not known where the two engineers lived but Robert Whitworth Jnr settled in Scotland and married Jane Fleming from Glasgow on 24 July 1793 at Old Kilpatrick, Dunbartonshire, near the western entrance of the Forth and Clyde Canal. Whitworth Jnr worked on other canals in Scotland, including the Union Canal with John Ainslie with its triptych of imposing viaducts. (He died in Scotland on 1 August 1807, only surviving his father by eight years.) Once difficult matters had been addressed with the construction, Robert Whitworth left his son in Scotland whilst he addressed other pressing tasks.

Bowling Locks, at the junction with the Forth and Clyde Canal and the River Clyde near Old Kilpatrick.

Port Dundas Upper Basin, Glasgow.

Lock below the Upper Basin, Dundas, Forth and Clyde, Glasgow.

Within four years the operation to deepen the Forth and Clyde Canal had been completed and arrangements had been put in place for the feeder to be built from the Monkland Canal. Jean Lindsay states that the finished canal was 38¾ miles long, 60ft wide at surface and 30ft wide at base. The summit from Stockingfield to Castlecary was 156ft above sea level with a rise of twenty locks to the east and nineteen locks to the west. Water for the western part of the canal was obtained chiefly from the Monkland reservoirs and from the east from the Townhead Reservoir. A new basin at Port Dundas, landscaped as a garden port, was planned and built with connecting roads to the city of Glasgow. The basin of the port was 900ft long and 100ft wide. There was a wharf measuring 50ft and the port was equipped with warehouses, granaries and timber houses. Ironically, the desires of the Edinburgh gentry became realised at Glasgow.

By June 1790, after the labours of 419 navvies, 130 masons, eighty-one quarrymen and twenty-five carpenters over four years, the new section of the canal opened from Bowling to Glasgow. The magnitude of the work and the spectacle of vessels navigating Kelvin Aqueduct, 400ft in length and 70ft high, led the *Scots Magazine* to claim that 'the canal had a pre-eminence over everything of a similar nature in Europe and does infinite honour to

Part of the redevelopment at Port Dundas, Glasgow.

Canal House, Dundas Basin.

the professional skills of that able engineer Robert Whitworth Esq., under whose direction the whole of this great work has been completed in a very masterly manner'.

During 1789 Robert Whitworth also surveyed a canal from the Forth and Clyde Canal to Bo'ness (Borrowstounness). The following year he surveyed a canal from Berwick-upon-Tweed to Kelso and Ancrum Bridge. Remarkably, while undertaking this major canal work in Scotland, Robert Whitworth found time to travel to Oxfordshire to give advice on the construction of Wormleighton Reservoir to the Oxford Canal Company and his old associate, Samuel Simcock. During this year, Whitworth also re-surveyed the Leeds and Liverpool Canal, recommending the summit level be deepened and a tunnel, 1500yds long, constructed at Foulridge. There are echoes here of the principles of water storage that Whitworth incorporated on the Forth and Clyde Canal. On the Leeds and Liverpool Canal he recommended the summit level be deepened to 7ft. During this period Robert also re-surveyed the Ashby Canal.

It is unclear where Sarah Whitworth was living at this time. Robert and Sarah had bought a house, Hood House near Burnley and the works of the Leeds and Liverpool Canal, so it is probable that Sarah and family remained based there whilst Robert did a very passable impression of his old master James Brindley; boxing the compass, pounding the turnpike roads and working on the country's canals, calling in to see family on his way back and forwards to Scotland. Father and son might well have worked alternately, supervising matters in Scotland as young Robert also worked on the Dearne and Dove Canal and surveyed a ship canal to serve Canterbury. Robert Whitworth Snr had also worked on the Dearne and Dove Canal, but by 1789 he had taken up position as the resident engineer to complete the Leeds and Liverpool Canal which, since 1770, had remained unfinished. Robert Jnr stayed to work in Scotland.

It is possible that at this time there were some family difficulties regarding money. Robert Snr and Sarah had seven children born between 1766 and 1781. Two of their

Canal House, Leeds and Liverpool Canal, Foulridge.

Gargrave Aqueduct over Eshton Brook.

sons, Robert and William, had followed their father into canal engineering. The eldest child Hannah had married Joseph Woodhead on 14 June 1787 in Halifax. Judging by comments in Robert Whitworth's Snr's will, he had lent his son-in-law Joseph Woodhead considerable sums of money. The purpose behind the loan is not clear but it weighed heavily on Robert Whitworth's mind twelve years later. In his will Robert left £40 to his granddaughter Sarah, born 1788, but £400 to her younger sister Mary, born 1790. He explained that this was because he had already lent or gifted considerable sums of money to her father Joseph Woodhead. By implication it seems the elder unmarried daughter had in some way been the recipient of money handed down through Whitworth. At the end of his will was the unusual caveat that anyone contesting or causing a disturbance over the will would forfeit their benefits which would be shared equally amongst the other beneficiaries. There is an interesting story behind this which cannot be unravelled at present.

Whitworth Snr's task was to complete the summit level of the Leeds and Liverpool Canal and join the two ends from Gargrave to Wigan. He was paid 600 guineas per annum; such were the fees that an engineer of his standing could command. Many contractors and labourers were brought south from Scotland to the works. The new, more heavily engineered, route through Burnley, Accrington, Blackburn and Chorley now meant that the canal travelled south through the more populated area of Lancashire. This pleased the Lancastrian group as this had been their original ambition.

In 1793 Robert Whitworth surveyed and costed the construction of the Dorset and Somerset Canal, designed to link Poole and Bristol. When construction commenced the canal was planned from Blandford Forum to the Kennet and Avon, via Wincanton and Frome, with a branch to Coleford and Nettlebridge. In the conclusion to his original report, Whitworth noted that he was too busy to undertake any more surveys but recommended his assistant William Bennett to the task and promised to support Bennett in this work. Whitworth was well placed to provide advice on the work, given his survey

Foulridge Tunnel, eastern portal, Leeds and Liverpool Canal.

for Brindley in 1769 for a link between the English and Bristol Channels. Only a small part of this canal was built.

By March 1795, Robert Whitworth, back in Yorkshire, had calculated the work needed to construct the Burnley Embankment which was 40ft high and 1 mile in length; one of the wonders of the British canal system. Most of the earth came from a cutting north of Burnley and was carried to the site in narrowboats. Spoil from the tunnel at Gannow was used for the other large embankment between Burnley and Hapton at Bentley Brook; a clear example of cut-and-fill in practice. This must have been a stable time in Whitworth's life as he was able to be based with Sarah in his own elegant house, Hood House, near the works in Burnley. Sadly this Georgian house was demolished in the 1930s.

Robert Whitworth was now one of the country's foremost canal engineers and was continually in demand for advice the length and breadth of the country. The work, with his own small school of engineers, namely his sons, made this possible. After preliminary advice and calculations he was able to leave his sons on site to supervise matters in his absence. This is illustrated by Robert Jnr's work in Scotland and William's work on the Leeds and Liverpool Canal. In 1795 Whitworth requested that the Leeds and Liverpool Canal Company halve his salary so that he could work on other projects; mainly those involving his sons. Both Roberts were at work on the Ashby Canal in 1794–95 until they were dismissed – the younger Whitworth being accused by the Ashby Company of spending too much time on other projects.

During the years 1793–95, Robert Whitworth returned to the South West surveying the Wiltshire and Berkshire Canal with his son, William. Robert Snr had walked part of this area at some time in 1788–89 with Samuel Weston when plans for the Western Canal, the forerunner to the Kennet and Avon Canal, were being mooted. After his father's death

Burnley Embankment, Leeds and Liverpool Canal.

Roving bridge near Gargrave, Leeds and Liverpool Canal.

in 1799, William continued with this engineering work. William, like Robert, settled next to his canal work, marrying Rebecca Court in 1809 in Oxford. By 1812 the couple were living with two of their children at Watchfield House near Farringdon. A third and final child, Julia, was born there in 1815.

By the mid-1790s Robert Whitworth Snr found himself in the same position as his mentor James Brindley had twenty years earlier. Many canal companies were beating a path to his door to request advice, or he was in demand to act as an arbitrator between companies and their contractors. From 1795–97, during the height of the 'Canal Mania', he became consultant engineer to eight different canal companies. These ranged from canals on the Welsh border, such as the Herefordshire and Gloucestershire Canal, to the Grand Junction Canal in the Midlands and South East. In 1796 he was arbitrating between the Lancaster Canal and their contractors in the North West, as well as the Glamorganshire Canal where his old canal school fellow, Thomas Dadford, was in bitter dispute with the Glamorganshire Canal Company over payments. In 1796 he was in dispute himself, involved in surveying the possible Commercial Canal from Chester to Ashby crossing the Trent and Mersey near Burslem; it was never built.

Between 1797 and 1798 Whitworth gave advice on four river navigations: the Trent, the Tyne, the Wear and the Don. He was also consulted about the Gloucester and Berkeley Canal and the Dudley Canal. In 1798 Whitworth described himself as 'an old superannuated engineer'. By then he had worked for over thirty years, often out in bone-chilling weather; a mixture of rain, sleet and wind soaking everything. On 20 February 1798 he made his will, but he was not to see the new century. Nearly a year later, on 30 March 1799, his death was reported in the *Newcastle Journal*. He died at the White Lion Inn in Halifax, still working. His cause of death was described as 'mortification of his foot brought on by wet and cold during the late severe storm'. The condition sounds as though he had contracted gangrene and died of blood poisoning. In reality, Robert Whitworth had also gone the way of his old master Brindley: overwork!

Robert Whitworth's will is illuminating in many ways. Chiefly as it reveals his love of learning; philosophy, mathematics and reading. Unlike Henshall who left a host of businesses, or Simcock with his farm, public house, land and cattle, Whitworth had left to 'his three sons all his library of books and his mathematical and philosophical instruments

Sowerby chapel and cemetery. Robert Whitworth was christened, married and buried in this area. The chapel has not survived and neither has Whitworth's tomb.

to be divided amongst them as near in point of value as any disinterested person to be appointed by my executors shall judge the same'. He also directed his executors 'to give to each of my daughters one set or sets of books of the value of fifteen guineas or thereabouts as they my daughters shall choose and as my executors think fit'.

Clearly the numerical size and value of his library and instruments were such that they were worth mentioning in the will. The rest of the will is standard; it leaves bequests in the form of annuities and lump-sums for his wife and an equal distribution of his estate between his six surviving children. What had happened with regard to his son-in-law Joseph Woodhead and his granddaughter Sarah is unknown, but this could be the only reason why he includes the phrase about disturbances and challenges to the will which will be handled by the executors 'causing the beneficiaries to loose their benefit'. Alternatively, he might have wondered how his two engineering sons might react to the bequests. One wonders why William, the younger son, was made executor and not Robert? Possibly because Robert was in Scotland, but then again William was several hundred miles away from Yorkshire in Wiltshire.

Robert Whitworth was buried at Sowerby chapel where he and most of his family had been baptised. The Leeds and Liverpool Canal Committee summed up their feelings of Whitworth's death in their minutes when they recorded that they 'had the greatest confidence in the established abilities and integrity of their deceased engineer and his particular attachment to their undertaking and that their declarations shall be made and entered in their proceedings as a tribute to his memory and of their esteem'.

Further praise is recorded by an obituarist in the *Newcastle Advertiser*, who wrote:

> He succeeded Mr Longbotham in the direction of that grand undertaking, the canal from Leeds to Liverpool; which will be to his memory a monument more durable than marble or brass. When his person, his private worth, his social virtues, and his tender affections shall no longer be remembered, this together with his important works upon the Clyde … and many other places will lastingly convey to posterity, with deserved celebrity, the extensive genius and scientific powers of this great and much regretted artist. He was extremely modest, unassuming and communicative; for what nature had denied him eloquence and colloquial rhetoric, she more than amply remunerated him in the lucid powers of his pen.

Of all the engineers from Brindley's school, Robert Whitworth had achieved most in developing methods for canal construction and river improvement. Although his talents are perhaps obscured by twentieth-century historians writing of the new breed of civil engineers – represented by Thomas Telford, William Jessop and John Rennie – Skempton and Wright both refer to Robert Whitworth as 'one of the most able engineers in England'.

Without Robert Whitworth's meticulous talents, James Brindley would not have seen his plans fulfilled. Whitworth not only completed the work Brindley had started with his Great Cross of canals but he went on to complete other significant cross-country waterways in England and Scotland, as well as improving the navigation of many of the feeder rivers and other ancillary canals during the period known as 'Canal Mania'. Robert Whitworth also provided an axis of advice to those engineers left after the death of James Brindley. He is one of our most important, eminent and neglected canal engineers.

5

JOSIAH CLOWES 1735–94

'I have heard a good character of him'
– John Smeaton, 1792

Josiah Clowes, engineer, surveyor contractor and canal trader, was born in 1735 in north Staffordshire. His parentage is uncertain but they were possibly William Clowes and Maria Whitlock, who married on 23 October 1718 in Stoke. Nothing is known about Josiah Clowes' early home life or education until he was sixteen. Evidence from the parish registers indicates that the Clowes families were related to families of yeomen and gentry who had lived in the area of Stoke for three centuries, before the Industrial Revolution brought Stoke and Josiah Clowes to prominence.

What is certain is that Josiah was the sixth and youngest child of the family, having three older sisters and two older brothers – including the third-oldest, William Clowes (1728–82). William Clowes married Jane Henshall of Newchapel on 1 January 1750. Jane was the older sister of Anne Henshall, who later married James Brindley. As no records exist of an early baptism for Josiah we may surmise that the Clowes family were initially non-conformists or Free Thinkers. However, Josiah and his two older brothers, William and John, were baptised into the Anglican faith on 9 July 1751 at Norton-in-the-Moors. This could indicate that the brothers were anxious to become approved establishment figures, given the Test and Corporation Acts, and make their way in engineering and business following William's marriage to Jane Henshall.

It is certain that William Clowes owned coal fields at Whitfield, Norton and Sneyd Green, to the north of Stoke-on-Trent. Josiah had interests in the mining concern of Samuel Perry and Co. at Sneyd Green where he met Charles Bagnall (1747–1814), a partner in the enterprise. It is equally probable that Josiah assisted his brother with his mining business and there learned knowledge of underground excavation, which was to prove so valuable in his later canal-construction work when tunnelling.

Clearly, the linking of the Clowes and Henshall families by marriage brought Josiah into the orbit of Hugh Henshall, who was assisting James Brindley in surveying the Trent and Mersey Canal in 1758, including laying out the line of the great tunnel at Harecastle. It is certain the two became close friends whilst Henshall was clerk of works to the new canal. We know from evidence submitted by Clowes to the committee investigating the feasibility of the Worcester and Birmingham Canal, on 24 May 1791, that Clowes was employed by James Brindley and worked with him from 1770 until the latter's death two years later. In answer to the committee's question, 'Are

St Bartholomew's church, Norton-in-the-Moors.

you an engineer?' Clowes replied, 'Yes, upwards of 20 years employed under the late Mr Brindley.'

By 1762 Josiah Clowes was established enough to contemplate matrimony, and on 30 December he married Charles Bagnall's sister Elizabeth, from Wolstanton. The marriage was by special licence. The bride was unable to sign the register and her mark was recorded. The wedding was an extended family affair with Hugh Henshall being a witness, a clear indication of the growing closeness of the two men. John Clowes, Josiah's elder brother, and the church warden were also witnesses; indeed the minister was a Jonathan Clowes. Tragically, Josiah's happiness was to be short-lived as Elizabeth Clowes died at Chell Heath, seven weeks and five days after the marriage. She was buried at Norton churchyard on 19 February 1763.

The premier engineering work in the kingdom, the Trent and Mersey Canal, gained its act of acceptance in 1766 and excavations began at Harecastle in July of the same year. Clowes undoubtedly had first-hand experience of the work, particularly when responsibility for the completion of the canal fell to Hugh Henshall following James Brindley's death in 1772. Josiah appears to have been acting in the role of contractor amongst other initiatives during these years. By 1773 it is possible that Josiah Clowes was working for the Staffordshire and Worcestershire Canal in some capacity. Had Brindley sent him there as part of the work within the school of engineers? Was he a contractor on this work? Was he there to advise on the tunnels, short though they were? We are not sure, what we do know is that he had remarried. His second wife's name was Margaret (her maiden name is not known at present). It would seem that he was in difficulties, possibly of a personal as well as a professional nature, as a letter from John Davenport to Thomas Clifford regarding business on the Staffordshire and Worcestershire Canal, dated 22 September 1773, reads:

Sir … I think Clowes shod. Have notice to quit next Spring As his circumstances are getting bad for which purposes I have sent a Copy of a notice to him of wich if you

THE CLOWES FAMILY

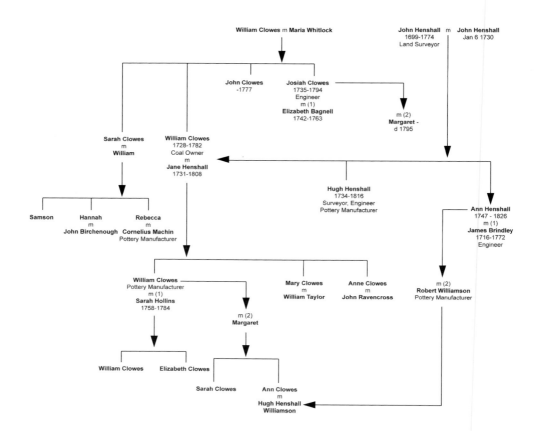

approve let two copies be made let Perry examine them, deliver one to Clowes or his wife and keep the other back whereof he shod. Make a memorandum of the day of the month when and on whom he served it, this must be done before 29th of this month or it will be too late – By Clowes removal you may have to accommodate both Bentley and Shaw.

In June 1775, whatever the problems, Clowes was back in the North Midlands, advertising from Middlewich for labourers in order to get the Trent and Mersey Canal completed. As early lock construction involved the laying down of a wooden framework in the excavated area prior to brickwork and masonry construction, Josiah may have been involved in engineering locks as well as the tunnel on the Trent and Mersey Canal, as he is later described, during his work on the Thames and Severn Canal, as resident engineer and carpenter.

The full nature of Josiah Clowes' work on the Trent and Mersey Canal, when assisting Hugh Henshall and James Brindley, can only be guessed at, as the records of the Trent

Narrow Bridge, showing the scale of the Trent and
Mersey Canal, Fradley Junction.

and Mersey Canal Company have been lost.
Could Clowes have been Henshall's assistant?
Did Clowes assist with the engineering of
the tunnels at Preston Brook, Saltersford and
Barnton? Josiah's interest certainly focused on
trade and business once the Trent and Mersey
Canal had been completed. Despite approaches
from the Chester Canal Company, in 1776
Josiah went into partnership with Hugh
Henshall in the latter's canal carrying company;
the official carriers for the Trent and Mersey
Canal. He worked from Middlewich in Chester,
now his home, and the proposed exchange port
for broad to narrow barges working between
Preston Brook and Stoke on the new canal. He
certainly invested in the expanding canal system
that he was now firmly part of by buying shares
in the Staffordshire and Worcestershire Canal.

In 1778 Josiah Clowes returned to engineering. In April of that year he finally agreed to
help the Chester Canal Company and replaced
Thomas Moon, clerk to the company, as 'general surveyor and overseer of the works'.
He was employed on an annual salary of £200. In return, he had to undertake 'to be
constantly resident in the neighbourhood of the works and at no time to absent himself
from the management without the leave of the committee in writing'. The Chester
Canal Company had clearly got wind of Clowes' various interests and peripatetic lifestyle
in the manner of his tutor, James Brindley, and his fellow engineers, with their valuable
and much sought-after knowledge of canal construction.

Between June and July, Clowes was engaged in repairing locks between Chester and
Beeston Brook, as leakage had prevented the passage of heavy barges. The following
month the company wrote to Josiah complaining that he had been absenting himself from
their works and threatening a stoppage of pay if this continued. In 1778 Clowes travelled
to Gloucestershire where he replaced another of Brindley's school of engineers, Thomas
Dadford, in building some of the locks on the Stroudwater Navigation. This work,
plus the pressure of his trading business in Middlewich, resulted in a rate of attendance
on site that was unacceptable to the Chester Canal Company. In October 1778, the
company ordered that 'Mr Clowes, the engineer be discharged for not attending the
works agreeable to contract'. For the next five years Clowes concentrated on his trading
and business interests on the Trent and Mersey Canal. In 1781 he occupied property in
Dog Lane, Middlewich, and in 1780, business being business, he hired two of his boats to
his former employers, the Chester Canal Company.

Above: Sapperton Tunnel, western portal, Thames and Severn Canal.

Left: Stroudwater Navigation, company's offices.

The year 1783 was critical for Josiah Clowes as it presented him with the opportunity to develop his skills and reputation as one of the principal canal engineers in the country. This chance was afforded by the Thames and Severn Canal Company, who appointed 'Mr Clowes of Middlewich' as 'surveyor, engineer and head carpenter' to their ambitious east–west canal link. Clowes' duties were to assist Mr Whitworth, the surveyor, in setting out the navigation. He was the resident engineer and was paid £300 per annum, although he became immediately the sole engineer for the enterprise.

This was one of the greatest tasks in canal engineering so far undertaken in Britain; that of linking the Thames and Severn and piercing the Cotswolds with the world's longest and largest canal tunnel at Sapperton: 3,817yds. Whitworth's initial survey did little other than mark out the line of the 9-mile summit level which included the mammoth tunnel. The project was then left in the hands of Josiah Clowes and James Perry, the salaried superintendent of the company who had no previous canal construction experience – although he had assisted with the finance and building of the Staffordshire and Worcestershire Canal and had probably met Clowes on that project. The canal company made clear the part, or lack of it, that Robert Whitworth played in the construction when they stated: 'Robert Whitworth was not at any time the only surveyor and engineer for the construction of the canal.' Clowes, now in his late forties, took up residence at Castle Street in Cirencester, near the works at Sapperton, and began what became an extremely difficult and onerous task.

The monumental tunnel at Sapperton had been marked out by Robert Whitworth in the company of Clowes. Miners were recruited from Somerset, Derbyshire and Cornwall to work on the tunnel. The general labourers on the canal were recruited locally from 'the populous vales of Stroud, Brimscombe and Chalford': this recruitment came as a blessing due to the depression in the clothing trade in the area. Tunnelling (as can be gleaned from the descriptions of the work in the second chapter) was an extremely dangerous and unpleasant task; Sapperton was no exception. The parish registers at St Kenelm's church, Sapperton, record the increased death rates in the village at this time, as well as the record of marriages of visiting miners to local girls. Between 1784–88 the record of deaths in the parish doubled given the high mortality rate in tunnel construction. A house was constructed at the eastern end of the tunnel which lodged the miners and canal workers. It can still be visited as a public house today, although it is not its original size as one of the floors was burnt in 1952.

The canal, like others, was constructed by the gang piecemeal system and not a single contractor; however, the miners comprised specialist work gangs. Here Clowes made the grievous mistake of employing one Charles Jones to build part of the tunnel. Jones had possibly worked on the Norwood Tunnel on the Chesterfield Canal with his Derbyshire miners. Jones was later described, after his son had threatened to murder one the company's employers and Jones had been imprisoned in Stroud Gaol for debt, as 'neither a skillfull artist, attentive to his business or honourable, but vain, shifty and artful in all his dealings'. Despite this testimonial, Jones went on to engineer the Greywell Tunnel on the Basingstoke Canal and later the Dudley Tunnel. Humphrey Household's verdict on Clowes was that his work lacked effective supervision. Yet sympathy must be extended to Clowes, working in professional isolation without Whitworth and without advice from Henshall or other members of the school of engineers.

In addition to the problem of rogue contractors, the geology of the tunnel itself presented difficulties. The tunnel passed through Fuller's earth, inferior oolite and great oolite; the bore moving through one set of unstable rocks to another. Leakage was the major area of concern with the tunnel during its construction and its subsequent use. In June 1789 Clowes confided his fears to the committee when he wrote: 'The tunnel loses water I myself heard it in Richard Jones's work … and I would recommend for all the rock parts that are not archd, to be very carefully examind, and sounded and get down those partes that are not sowned and firm immediately.'

In another letter, dated 26 July 1789, Clowes states with regard to the tunnel, 'I have done all that I could to discover aney errors in the dangers of my life maney times'. Despite Clowes' attempts to seal the tunnel with 'oak

St Kenelm's church, Sapperton, where the navvies are buried.

Above: Tunnel House, Hailey Wood, Thames and Severn Canal.

Left: Josiah Clowes' inscription at the Red Lion Lock, Chalford, Thames and Severn Canal.

noggs in the scafold holes, re-puddling to the correct height, racking and gravelling the invert and re-bricking part of the Long Arching', the real error in the tunnel related to the properties of Fuller's earth which swelled when wet. The consequent movement of soil caused significant leakage in the invert but no eighteenth-century engineer had knowledge of the action and properties of this soil. There were also problems with the underground springs that Robert Mylne attempted to resolve in 1790.

In July 1789, Clowes wrote plaintively to James Black, canal proprietor: 'Now I wish you would send me a man that understands canal navigations and then he may inform the gentlemen [the committee] his opinions.' Nevertheless, in 1820 Baron Dupin described Sapperton Tunnel as 'one of the finest specimens of its kind', and in 1870 the then manager of the canal, John Taunton, remarked that 'the engineering seems to have been of a perfect character'. On the canal's completion *The Times* thundered that

Sapperton was 'The grandest object ever attained by Inland Navigation'. At Red Lion Lock in Chalford, Clowes celebrated more modestly by having his name inscribed briefly over the keystone of the masonry lock bridge. Elsewhere, events were more ominous for British trade and power: George Washington had been elected the first American president and the Bastille had been stormed.

The work, including the tunnel, was eventually completed in five and a half years; half the time it had taken to build Harecastle, despite being 837yds longer and 117sq. ft greater in cross section. (Whitworth had been uncertain of the size of barges to navigate the tunnel as the standard narrowboat had not been adopted by the company.) In addition, between 1789 and 1790, Clowes supervised the building of locks on the Upper Thames Navigation at Osney, Buscot and St John's; this in order to improve the passage of boats from Inglesham to Oxford and on to London.

There is reference during the construction of the tunnel to 'Mr Clowes's Model or Driving Frame'. No patent exists in his name for such a device and it is unclear what the machine consisted of. It is an intriguing speculation that it was the forerunner of Marc Brunel's tunnelling shield, but it was more likely to have been a frame for measuring the dimensions within the tunnel. Clowes was a believer in new technology, however, as exemplified by his use of John Carne's cutting machine, and he had a timber railway installed in the great tunnel to withdraw spoil.

A pleasing break from Clowes' political and engineering headaches came on 20 April 1789 when King George lll visited the tunnel works at Sapperton. The king came whilst he was taking the waters in Cheltenham for medicinal purposes. Later that year he was to suffer the first of his descents into madness due to the onset of porphyria, but clearly the opportunity to visit the greatest engineering work in the kingdom came as an inspiration to the king, and a welcome recognition of his efforts for the harassed Clowes. There were also always the other awkward political tasks that the canal company got its resident engineer to complete. The Revd Keble had complained of the canal flooding his meadows. Clowes wrote a memorandum on 24 November 1787:

> By order of the Committee I went to view Mr Keble's meadows, and the culverts made to draw off their flood waters and I find that more water might go than is necessary and that these meadows are not injured by the stoppage of any floodwater thro the deficiency of the culverts.

Despite the many difficulties recorded in the building of the canal, and in particular the tunnel at Sapperton, Clowes left the work of the Thames and Severn Canal with his reputation enhanced, gaining good reports from both Whitworth and Smeaton. Smeaton wrote to the Worcester and Birmingham Canal Company in 1792: 'Mr Clowes I do not personally know, he was employed upon the Thames and Severn Canal … I have heard a good character of him but whether he was employed in any principal work before that is unknown to me.' From 1789 onwards, during the last five years of Josiah's life, he became much sought after by canal companies.

By 1789 'Canal Mania' had gripped the country and proprietors clamoured for experienced engineers and surveyors. Where tunnelling was required, Josiah Clowes' expertise and advice was sought. That year Clowes left the Thames and Severn Canal to complete the Dudley Tunnel (3,172yds) and remedy the errors in its alignment. The

Dudley Tunnel, southern portal, on the canal's reopening in 1973.

Dudley Tunnel, northern portal.

Dudley Canal and its tunnel were built to facilitate the easy transportation of coal from Viscount Dudley's mines at Tipton to the Thames and Severn Canal, via the Stourbridge and the Staffordshire and Worcestershire canals. Clowes received one and a half guineas a day for his advice from the Dudley Canal Company and all his expenses were paid.

Clowes was now becoming a prosperous man and to underline the point he bought property in Marsh Green in Biddulph, and Lewin Street in Middlewich. After two years Josiah completed the work at Dudley started by Thomas Dadford Snr; the tunnel was opened throughout and the reservoir at Gads Green finished.

The year 1791 was an extremely busy one for Josiah Clowes. He gave evidence to the Commons on the Leominster Canal, the Worcester and Birmingham Canal and the Upper Thames Navigation. He was also involved in surveying and engineering the Herefordshire and Gloucestershire Canal – 35 miles 5 furlongs in length. Clowes authorised the gin wheels from Sapperton Tunnel to be sent to Oxenhall Tunnel to assist with this work. John Carne's cutting machine from the Stratford Canal was also requisitioned. However, the Oxenhall Tunnel (1320yds) and the canal were not opened until 1798, four years after Clowes' death. Its construction was completed by Robert Whitworth. These canals were designed to link to the Severn, the Staffordshire and Worcestershire Canal and the Thames and Severn Canal – Clowes' first canal.

The travelling that Clowes undertook throughout the country in execution of his canal work was still hazardous, despite the improvements to turnpike roads. R.R. Angerstein wrote:

> the nearer we got to London, the more travellers were seen on the highway. The latter were generally anxious to find some company in view of the danger of being robbed by highwaymen, as so often reported in the London Gazette. Many reminders of this were seen hanging in gallows along the highway.

Oxenhall Tunnel, Herefordshire and Gloucestershire Canal.

To underline the authority that Clowes was regarded with within engineering circles and by canal companies, particularly when tunnelling was involved, he was invited in 1791 to inspect the building of the Leeds and Liverpool Company's summit tunnel at Foulridge. The company suspected that Robert Whitworth had committed errors in its construction. Josiah indicated a shorter tunnel by 600 or 700yds could have been dug, but concluded it would be more sensible to complete Whitworth's line as planned. On 11 February 1791, Clowes had already surveyed the summit level and been paid 20 guineas for doing so.

By January 1792, the continued blandishments of the Worcester and Birmingham Canal Company eventually resulted in Josiah Clowes accepting the position of consultant to the company for 2 guineas a day. Canal work was now becoming highly remunerative. Josiah checked John Snape's survey and also became employed to survey and engineer the Stratford and Dudley No.2 canal, with John Snape as assistant. The latter canal project was tied to the development of a more direct route from Birmingham and the Black Country coalfields to the Severn at Worcester, which was needed to circumvent the heavy tolls of the Birmingham Canal Navigations. On the Stratford Canal, Clowes engineered the Brandwood Tunnel and on the Dudley No.2 Canal tunnels again dominated. The entire Dudley No.2 line was only 13 miles long, but over 5 miles were in tunnel at Gosty Hill and Lapal. The later was the most ambitious, where a 3,795yd tunnel was constructed under the California coal mines. Clowes did not live to see the line completed and the aptly named William Underhill took over the enterprise with Thomas Green after Clowes' death.

The Lapal Tunnel, like its forebears, presented many problems. Shortly before his death, Clowes published a report for the Dudley canal company which included information of the dimensions of the tunnel and the brickwork, in some cases only one brick thick. Later there was to be an issue concerning the dimension of the tunnel and the ability

Foulridge Tunnel on the Leeds and Liverpool Canal.

Right: Brandwood Tunnel, Stratford Canal.

Below: Gosty Tunnel, Dudley No.2.

of the side walls to withstand the pressure exerted by groundwater, fluid marl and clay. The tunnel passed through faults and there were subterranean springs similar to its predecessor at Sapperton. Here, thirty shafts were needed for construction and 500,000 bricks were made at Moor Street. Eventually, three steam engines were purchased to drain the workings.

By now Clowes was exhausted and possibly ill. William Underhill completed the work but the canal company were not impressed with his capabilities and, emphasising their classical knowledge, wrote:

> Our engineer has been very unequal to the undertaking in the point of judgement and conduct, but now that he has had the advantages of experience, and is perfectly acquainted with the whole of the works, it is thought more prudent to continue him under the direction of the Clerk of Works; who have had an Augean Stable to cleanse.

In 1798 the tunnel was completed and William Underhill died in Cheltenham in 1804.

Between 1792 and 1794 Josiah Clowes was working for six canal companies. Again in 1792, Clowes surveyed the line of the Gloucester and Berkeley Canal and estimated its building cost at £102,108. This canal was engineered by Robert Mylne, with James Dadford as resident engineer from 1795. Clowes continued to work as surveyor and engineer to the

Herefordshire and Gloucestershire Canal whilst undertaking the survey and engineering of the Shrewsbury Canal. His survey and costing of the latter was achieved in six months and were completed by 1 January 1794. On the Shrewsbury Canal he tunnelled again at Berwick (970yds); the first tunnel of any length to be built with a wooden cantilevered

Left: Junction of the Stroudwater and the Gloucester and Berkeley Canal.

Below: Berwick Tunnel, Shrewsbury Canal.

towpath. This addition was suggested to Clowes by William Reynolds, the iron master. Other features of the canal included guillotine gates to conserve water and the unique cast-iron aqueduct at Longdon-on-Tern. The latter, however, was built by Thomas Telford after the central arches of Clowes' original aqueduct were washed away by floods in 1795. In April 1794 Clowes returned to the Stroudwater Navigation to report on improvements to be made to the canal entrance with the Severn at Framilode.

Josiah Clowes died at Middlewich during late December 1794. Josiah Wedgwood's valediction to James Brindley might very well apply to Josiah Clowes. Wedgwood commented: 'I think Mr Brindley – the great, the fortunate, money getting Brindley, an object of pity ... he may get a few thousand but what does he give in exchange? His health, and I fear his life too.' Clowes did die a relatively wealthy man with shares in excess of £1,000 in many of the companies he was working for. He also owned several properties in Cheshire and Staffordshire. These, plus his linen, plate, china and clothes, he left to his second wife, Margaret. Margaret Clowes only survived Josiah by months and, as there were no children, his nephew William Clowes, son of William and Jane Clowes, *née* Henshall, inherited his estate.

Josiah Clowes' remains were taken from Middlewich and, at his request, as recorded in his will, they were interred in Norton-in-the-Moors churchyard,

1 Canal boats at the Gas Street basin
(once known as Brickkiln Place),
Birmingham Canal Navigations.

2 The tomb of John Brindley, James'
brother, St Peter's church, Kinver.

3 Chesterfield Canal, near Worksop.

4 Oxford Canal at Napton Hill.

5 Stroudwater Navigation, west of Stroud, before restoration.

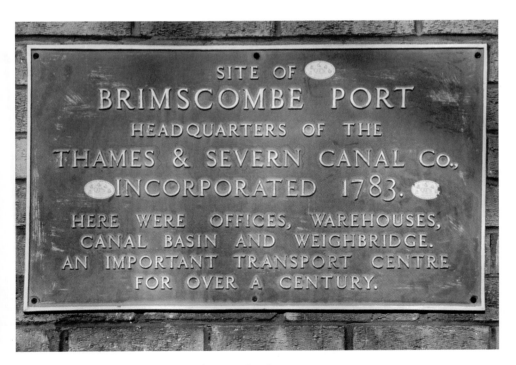

SITE OF
BRIMSCOMBE PORT
HEADQUARTERS OF THE
THAMES & SEVERN CANAL Co.,
INCORPORATED 1783.
HERE WERE OFFICES, WAREHOUSES,
CANAL BASIN AND WEIGHBRIDGE.
AN IMPORTANT TRANSPORT CENTRE
FOR OVER A CENTURY.

6 Plaque, Brimscombe Port, Thames and Severn Canal.

7 Sapperton Tunnel, eastern portal at Coates.

8 Wiltshire and Berkshire Canal, Dauntsey.

9 Wharves at Bowling, Forth and Clyde Canal.

10 Kelvin Aqueduct, Glasgow, Forth and Clyde Canal.

11 The Birmingham Canal at Aldersley Junction.

12 Leeds and Liverpool Canal near Gargrave.

13 Froghall Wharf, Caldon Canal.

14 Red Lion Inn, Ipstones, where Brindley was placed into a damp bed.

15 Wilts and Berks Canal: preserved section plus new lifting bridge near Foxham.

16 St Mary's, Ardley, where Samuel and Esther Simcock are buried.

17 Oxford Canal meanders near Somerton.

Hadley Park, Guillotine Lock Gates, Shrewsbury Canal.

Longdon-on-Tern Aqueduct, Shrewsbury Canal.

Josiah Clowes' grave, where he is buried with his wife, St Bartholomew's church, Norton-in-the-Moors.

View of Caldon Canal from the road leading to St Bartholomew's church at Norton-in-the-Moors. Here Josiah Clowes was baptised and is buried.

Grave of Williams Clowes and his wife Jane Clowes, *née* Henshall, at Norton-in-the-Moors.

next to those of his short-lived first wife Elizabeth, on the morning of New Year's Day 1795. His second wife, Margaret, died aged fifty-three, four months after him. She was interred in the same tomb on 22 March 1795. The Georgian church overlooks the pastoral Churnet Valley and the Caldon Canal on one side, and the industrial landscape of Hanley and Stoke-on-Trent on the other. Both were shaped, to a degree, by the engineering talents of Josiah Clowes and the Brindley school of engineering.

William Clowes, Josiah's nephew, put his inherited wealth to good purpose. By 1781 he had founded the Newhall pottery company with Jacob Warburton, Samuel Hollins and Charles Bagnall. It was a successful ceramic business which bought up Champion's patent for the manufacture of true porcelain. Eventually William Clowes secured the controlling interest in this company, as well as founding his own pottery works at Longport with Hugh Henshall and Robert Williamson. A pleasing postscript to Josiah's life lies in the marriage of Anne Clowes, his great-niece, to Hugh Henshall Williamson, one of the sons born to Anne Brindley after her second marriage to Robert Williamson. The Clowes, Henshall and Brindley families united, having been brought together by the world of canals and pottery.

There is no known portrait of Josiah Clowes and no personal letters appear to have survived. All that exists is professional correspondence, a map of the Herefordshire and Gloucestershire Canal and his will. Nothing is known, either, about his early education. However it is likely, given Josiah's ability to write reasonable English and to undertake the mathematical and scientific skills required for surveying and engineering work, that he received some schooling, either from a dame school or from his family and friends. We also have Smeaton's report on his 'good character' and this is likely to be a realistic appraisal as Hugh Henshall, his close friend and associate, was regarded as very hard-working and a man of the highest integrity.

Spanning the era of twenty years from James Brindley to Thomas Telford, Josiah Clowes' true importance as a canal engineer has not emerged in canal histories written in the last two centuries, and this despite Charles Hadfield referring to him as a 'canal veteran' in *Waterways to Stratford*. The majority of Josiah's canals in the North, West Midlands and the Welsh borders were planned and built with a vigour that seems astonishing considering that most of his work occurred in the last six years of his life. It is likely that some of his work was innovatory, given references to his model or driving frame on the Thames and Severn Canal – although this cannot be proved. Clearly, Clowes was keen to utilise new technology as exemplified by his use of John Carne's canal cutting machine on the Herefordshire and Gloucestershire Canal. Josiah Clowes, like many other engineers, was certainly guilty of undertaking too many projects at the same time during the height of the canal mania, but his undoubted contribution to the country's civil engineering development lay in his achievement as England's leading canal tunnel engineer during the decade starting in 1784.

By the end of the canal-building era, Josiah Clowes had been responsible for the engineering or surveying of the second, third and fourth-longest tunnels on the English canal network. It also seems probable that he played a part in assisting with the construction of the country's first, great canal tunnel at Harecastle. Joseph Priestley asserts that Clowes was consulted over the building of the country's longest tunnel at Standedge (5,456yds) on the Huddersfield Narrow Canal, but no evidence for this exists in the company's minute books.

Josiah Clowes life's work is all the more impressive when one considers that he was partly self-taught in his chosen profession; learning engineering, surveying and tunnelling skills as he worked under the tutelage of James Brindley, Hugh Henshall and Robert Whitworth in their school of canal engineering. He was a surveyor, engineer, contractor, carpenter, trader and carrier, and left a legacy in civil engineering work in England that must have influenced his successors in both canal and railway tunnel work. Much of Clowes' engineering work is still to be seen in parts of the country today. The fact that he rose successfully to the top of his chosen civil engineering profession is a tribute to his energy, character and belief. On his tomb he is remembered with the simple inscription 'Engineer'; an interesting contrast to Henshall who claimed the title 'Esquire'. Clowes built the canals that Henshall, Whitworth and Brindley surveyed and planned, as well as extending the system further through his own surveying and engineering skills.

6

Thoˢ: Dadford

THOMAS DADFORD _c.1730–1809_

'Midwives of a new age'
– Charles Hadfield

Carpenter, builder, surveyor, engineer and architect; Thomas Dadford was an important member of James Brindley's school of engineers. No written proof exists that they met in person but it seems very unlikely they did not whilst both of them were working on the Staffordshire and Worcestershire Canal. However, their working lives were for the most part conducted at a distance.

It is probable that Thomas Dadford was born in Wolverhampton or its surrounds in 1730, although his parentage, baptism and birth date are unknown. His age is deduced from information given at the time of his marriage. It is possible that his mother was Priscilla Dadford whose burial is recorded in the Wolverhampton registers on 17 April 1769, and it is almost certain that he had a sister, Elizabeth, who married Charles Alton in Wolverhampton in November 1758. A Thomas Dadford was a witness at this wedding and it seems too big a coincidence that they were not the same person.

Also unknown is information about his early life, schooling and work until 1759. Evidently he must have received a reasonable education with an emphasis on mathematics and natural science. This equipped him to become a carpenter and builder in the first instance, although, like James Brindley, his spelling was persistently phonetic. John Norris, in his extensive researches on the Dadfords, opines that he was recruited to work by the Staffordshire and Worcestershire Canal Company and was approved by James Brindley because of his established reputation as a carpenter and builder in that area. This seems likely.

Our first recorded knowledge of Thomas Dadford comes on 9 September 1759 when he married Frances Brown, daughter of Samuel Brown, a toymaker in Wolverhampton. (A toy, in the eighteenth century, was a small fashion article or trinket.) The marriage took place in the handsome church of St Peter's in Wolverhampton and was by special licence: the couple were Roman Catholics. Wolverhampton and the area of the West Midlands had a positive reputation for the toleration of the Catholic faith. Giffard House, built between 1727 and 1733 in Wolverhampton, was the first public Mass House and priest's residence in England which was in continuous use.

The couple had four sons and a daughter, Mary. Three of the sons, Thomas Jnr, James and John Dadford, followed their father into canal building and worked closely with him during the period of the 'Canal Mania', particularly in Wales and the south-west

St Peter's church, Wolverhampton, where Thomas Dadford married Frances Brown and where they were buried.

Midlands. The youngest son, William, went to America. Norris thinks this was to search for his older brother John, who had gone to America in 1796. Whether William was involved in engineering there or in Britain is not known at this stage.

Following the famous meeting at the Wolseley Bridge Inn in December 1765 to set out the Navigation between the rivers Trent and Mersey, there was a separate meeting, a month later on 20 January 1766. Led by James Perry, a businessman from Wolverhampton, this meeting sought to join the Severn by another canal to the newly planned Trent and Mersey Canal. This was the Staffordshire and Worcestershire Canal. Hugh Henshall and Samuel Simcock carried out a preliminary survey with James Brindley, who had the final word on the detailed laying out of the line. The acts of royal assent for the Trent and Mersey and the Staffordshire and Worcestershire canals were granted on the same day, 14 May 1766.

46½ miles long, the Staffordshire and Worcestershire Canal left the Trent and Mersey Canal at Great Haywood and rose to a summit at Gailey by way of twelve locks. After the summit level of 10½ miles the canal then fell through thirty-one locks to Stourport and the Severn. There were two short tunnels on the line at Cookley and Stourton. The dimensions of the locks were identical to the Trent and Mersey Canal's, so that the gauge was standardised on the canals of the Great Cross. Indeed, it became the first part of the

THE DADFORD FAMILY

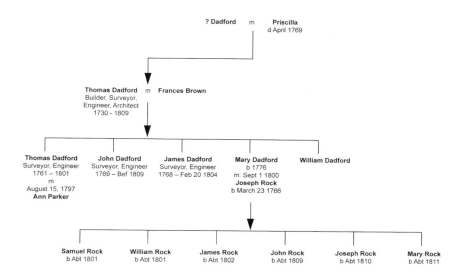

```
                    ? Dadford    m    Priscilla
                                      d April 1769

        Thomas Dadford    m    Frances Brown
        Builder, Surveyor,
        Engineer, Architect
          1730 - 1809

Thomas Dadford      John Dadford       James Dadford      Mary Dadford       William Dadford
Surveyor, Engineer  Surveyor, Engineer Surveyor, Engineer   b 1776
 1761 – 1801         1769 – Bef 1809    1768 – Feb 20 1804 m: Sept 1 1800
      m                                                    Joseph Rock
August 15, 1797                                            b March 23 1766
  Ann Parker

  Samuel Rock      William Rock       James Rock       John Rock        Joseph Rock        Mary Rock
  b Abt 1801       b Abt 1801         b Abt 1802       b Abt 1809       b Abt 1810         b Abt 1811
```

Cross to be completed in 1772. After the initial survey Brindley retained oversight as consulting engineer and was required by the company to visit weekly to check on progress and issue instructions. His assistants at the canal face were Samuel Simcock and Thomas Dadford. The clerk of works was John Baker, although Peter Cross Rudkin points out that Baker was not an engineer and the work was done by John Green, the under-clerk of the works.

On 17 March 1767, Thomas Dadford was appointed as carpenter and joiner to the canal company, 'to serve us in the way of his said Trades in providing and carrying on this Navigation for the space of Five Years if the said Navigation shall not be sooner Compleated at a Salary of Seventy pounds pr. Annum'. James Brindley instructed that it was 'Mr Dadford's sole employ to attend to the building of lockes'. Whether this was because Dadford was acquiring

Dunsley Tunnel near Kinver, Staffordshire and Worcestershire Canal.

Compton Lock and
bridge, Staffordshire and
Worcestershire Canal, the first
lock built on the Brindley system
according to Joseph Priestley, canal
historian and son-in-law of Robert
Whitworth.

Compton Lock, Staffordshire and
Worcestershire Canal.

great expertise as a lock engineer or whether it was because he may have been trying to undertake other work not allocated to him is not clear. Norris suspects the latter. According to Joseph Priestley, one of the first locks constructed on the Brindley canal system was built at Compton on the Staffordshire and Worcestershire Canal, and Dadford was presumably supervised by Brindley himself. On 23 July 1767 Dadford set out the triple staircase of locks at Bratch. Later, the locks were to be altered with side pounds between the chambers. James Brindley had been experimenting with lock design (see the chapter on Brindley for his possible experimental work at Turnhurst Hall). If this did not take place then, he would have most certainly have experience in the construction of broad locks from his work on the Calder and Hebble Navigation.

Thomas Dadford was also responsible for one other significant masonry structure on the canal. In 1771 he constructed a substantial aqueduct across the River Sow near Shugborough. This work was undertaken under a separate contract from the Staffordshire and Worcestershire Canal Company. In 1774 Elizabeth Prowse, accompanying a boat trip made for pleasure by the Sharpe family, comments in her itinerary of the voyage that the locks at Stourport had been engineered by Thomas Dadford. She writes: 'Kidderminster Tunal under the Street 120 feet. Storeport a Basson that covers 3 Acres and locks down into the Severen. great Works done in 5 years by Mr Datford, and within 12 M of Woster.' Elizabeth's spelling was even more imaginative than Brindley's and his colleague, Dadford.

In 1772 the Staffordshire and Worcestershire Canal was opened throughout and it is clear, from Peter Cross Rudkin's detailed paper on the construction of the canal for the *Newcomen Society Transactions*, that on 22 September 1773 a letter from John Dadford to Thomas Clifford revealed: 'Dadford is appointed principal engineer with a salary of £100 per annum wich Baker is apparently much disgusted.' Whatever Baker's feelings, Dadford was now resident engineer of one of the principal and most successful canals, in terms of

Sow Aqueduct, sometimes called Milford Aqueduct, built by special contract by Thomas Dadford on the Staffordshire and Worcestershire Canal.

Stourport Basin, Staffordshire and Worcestershire Canal.

use and profit, in England. In 1774 he, and presumably his family, lived at Compton. The Staffordshire and Worcestershire order book of 1766–85 records that on 21 April 1784 'a Dwelling House be built at Compton under the direction of the messrs C. Perry Lewis and Company … with all convenient speed for the habitation of the said Thos. Dadford.'

After several unsuccessful proposals in the early part of the eighteenth century to link the important cloth-producing town of Stroud to the Severn by a waterway, the possibility of such a navigation was raised again in 1774 following the completion of Staffordshire and Worcestershire Canal. The economic argument for cheaper canal transport was now unanswerable as long as the mill owners could be persuaded they would not lose water to power their factories. John Priddy, who had successfully engineered the Droitwich Canal following Brindley's survey, mapped a possible route in 1776 with Thomas Dadford at an estimated cost of £16,750. The new route avoided the mills and included conventional locks which Dadford was now becoming a master in constructing. In 1755 Thomas Bridge from Tewkesbury had sought to build the Stroudwater Navigation without locks; substituting instead crane transhipment points for goods where the levels changed. This was in order to placate the mill owners who were in dispute over water supplies. The canal, an early type of containerisation, was never built.

Thomas Dadford now turned his attention to significant canals feeding the Staffordshire and Worcestershire Canal further north. The canals concerned were the Stourbridge and Dudley canals. In 1776 Dadford became surveyor and then engineer to the Stourbridge Canal. James Brindley had surveyed a branch from Stourbridge to the Staffordshire and Worcestershire Canal in 1766, and this line had been amended nine years later by Robert Whitworth. These surveys were undertaken so that the abundant mineral resources— coal, ironstone, limestone and fireclay – lying in great quantities at Sedgley and Dudley could be tapped and transported by the Staffordshire and Worcestershire Canal. On 6 June 1776, the first meeting of the Dudley Canal Company was held at the Swan Inn in Dudley. The Stourbridge and Dudley canals had also gained their act of royal assent on the same day in 1776.

In his proposal to work for the Stourbridge Canal Company, Dadford said, with the assurance of a good salesman, 'Exparance as taught me every Imperfection of locks; therefore I should make your Locks mutch improved and to continue with little Repares or Indarence to your canal …'. Dadford's eldest son, Thomas Dadford Jnr, assisted with the works on the Stourbridge Canal until told to desist by the canal company. However, Dadford Snr's expertise with lock construction were much needed; as exemplified by the flight of sixteen locks at Stourbridge and later nine locks on the Dudley Canal. This, plus the four Parkhead locks, takes the Dudley Canal to the level of the Birmingham Canal at Tipton.

The building of the locks on the Stourbridge and Dudley canals further strengthened Dadford's knowledge of engineering canal structures and, in the years from 1770–81 as John Norris' detailed researches reveal, Dadford had acquired skills as an architect as well as a builder. In 1775 a Mr Dadford, well known in Roman Catholic circles in the West Midlands, was invited to submit an estimate for the rebuilding of the tower and spire at St Cassian's church, Chaddesley Corbett. John Roper, who wrote a history of the church, thought that Dadford had been approached as a result of intervention by the Throckmorton family. In the event Dadford did not rebuild the tower. This was left to James Rose, a local stonemason and architect, but it does give insights into Dadford Snr's developing building abilities.

St Cassian's church, Chaddesley Corbett. Thomas Dadford was asked to redesign the church and his eldest son, Thomas, was married here.

In 1776 and 1778 Dadford was invited to report again on the Stroudwater Navigation. Dadford undertook this survey and report, drawing attention to poor workmanship and inaccurate levels with the locks, but he was too busy with canals in the Midlands to be actively involved in the remedial engineering and recommended Josiah Clowes to examine the issue of the locks. By this date Clowes, who had started his canal career as a carpenter, had also acquired a reputation as a successful builder of locks. The Stroudwater Navigation was eventually completed in 1779 but Clowes was not available to work exclusively on this project.

A few years earlier, in 1775, the work on the construction of the Stourbridge and Dudley canals gave impetus to a plan to join these canals, and the Staffordshire and Worcestershire Canal, to the Viscount Dudley and Ward's existing tunnel and mines under Castle Hill to the north. This would release minerals directly to Birmingham. Some time elapsed before this proposal to link the underground canals, under the high ridge of ground at Castle Hill at Dudley, to the Dudley and Stourbridge canals took place. John Snape (1737–1816), a land surveyor from Warwickshire, carried out a survey of this tunnel extension and estimated the cost at £18,000. The tunnel planned was 14ft high, 9ft wide and had a depth of water of 5ft. Royal assent was granted on 5 May 1785. By September an advertisement appeared in *Aris's Birmingham Gazette* for someone 'willing to undertake the Execution of the said Tunnel'. It was at this point that Thomas Dadford decided to offer to oversee the project; it was to be one of the greatest of the canal structures he had been involved with so far. He agreed to act as a consultant engineer, supervising the construction of the great tunnel – at the time the third longest on the British canal system.

John Pinkerton, a well-known contractor from a family of canal builders, had worked on the Birmingham Canal and became the contractor for the Dudley Tunnel. John Wildgoose was appointed surveyor under Dadford to superintend the tunnel. Abraham Lees also assisted Thomas Dadford with the engineering. After the initial marking-out of the tunnel and an additional survey, Dadford recommended the tunnel be lengthened by 30yds; thereby reducing the deep cutting planned at the southern portal at Parkhead. He also suggested that the four locks bringing the Dudley Canal up to the level of the tunnel be increased to five; this alteration to save water. The canal company carried out these proposals but mining subsidence caused problems and, a century later, the Parkhead flight was reduced to four locks again. Dadford also extended the dimensions of the tunnel slightly and deepened the canal by 6in in order to provide additional water supply.

Whilst good progress was made at the Parkhead end of the tunnel, the same was not true of the northern end where many problems existed with the geology and the contractors wrestling with it. Dadford prepared a plan and estimate for a basin at the north end of the tunnel to meet with Lord Dudley's tunnel, but this was his last work for the company. In 1786 Dadford decided to leave the Dudley Canal Company after ten years of work. He had been offered a more lucrative position with the Trent and Mersey Canal as their resident engineer. It should be noted that Dadford had already worked in that area as he had been appointed surveyor of bridges to Staffordshire's Quarter Sessions in 1785; a job he held until 1792. He had also completed surveys for the Trent and Mersey Canal in 1781. Isaac Pratt did not succeed with the work on the tunnel and in 1792 Josiah Clowes, following in Dadford's footsteps again, eventually completed the tunnel at Dudley.

In 1781 Thomas Dadford Snr and his son Thomas surveyed the River Trent for the Trent and Mersey Canal Company, in order to render improvements to the navigation.

This they did, with estimates of cost, by 22 November of that year. Hadfield and Skempton, in their book on William Jessop, report that the Dadfords 'recommended a horse towpath and some side-cuts, weirs and locks to take craft of 40-50 tons at the cost of £20,000'. A Bill was introduced in 1782 but withdrawn, and it was left to William Jessop with John Pinkerton to undertake this work which was legislated for in 1783.

Dadford and his wife now moved to Staffordshire, but may have kept their property or properties in the Wolverhampton or Worcester area as they were to return there later. Their eldest son Thomas had married Ann Parker of Bluntington Green, Chaddesley Corbett, on 15 August 1797; Mary Rock, Thomas' sister, was a witness. Both partners were Roman Catholics but there were no children from their union. Norris suggests that by the time of this late marriage Thomas Jnr was possibly ill, and needed a nurse rather than a wife in the more conventional sense.

As the records of the Trent and Mersey Canal have been lost, it is impossible to evaluate the nature of Thomas Dadford's work and its effectiveness. However, given that Dadford remained in post for seven years on the premier canal in the kingdom and left to pursue other engineering interests of his own device, his service must have been of good quality.

Dadford and his wife next moved to the North Midlands at Corbridge, and here Dadford would have certainly met of one of the earliest of Brindley's school of engineers, Hugh Henshall. Henshall was by now running his carrier service on this canal as well as having interest in a multitude of local businesses, including pottery and mining. The two were living close to one another and Josiah Clowes, who Dadford also knew, had lands and properties in close proximity.

In 1788 Thomas Dadford Snr wrote to John Sparrow, the clerk to the Trent and Mersey Canal, to inform him that he intended to visit John Powell at Abergavenny. The Brecknock and Abergavenny Canal was planned to link Abergavenny to the Monmouthshire Canal and so to Newport. It was being engineered by two of Dadford's sons, Thomas and John. It is possible he was being called upon to advise on surveying or lock and aqueduct construction. However, Powell was also concerned with the development of the Glamorganshire Canal and it is equally possible that Dadford Snr was thinking ahead, with his sons, as to further engineering opportunities.

Two years later his engineering achievements were nationally recognised by the Smeatonian Society when, on 25 April 1783, 'Mr Dadford of Corbridge near Newcastle under Lyne was unanimously elected a member of this society.' A year later Thomas Dadford was working with Robert Whitworth on the proposed aqueduct over the River Tame on the Coventry Canal. As they met at High Holborn at the Smeatonian meeting, it is probable they discussed this work over refreshments in London. Brindley's work on the Great Cross was now being completed by all of his chief students.

Between 1785–87 Dadford was involved in engineering work on the Fradley–Whittington Brook section of the Trent and Mersey Canal, and, from 1785, Dadford was working again on the Coventry Canal surveying the Fazeley–Fradley Junction section. He had now worked on nearly all branches of the Great Cross, with the exception of the Oxford Canal. It was during this period that Dadford met Thomas Sheasby (1749–99). Sheasby came from Tamworth and was a contractor and engineer working on the Birmingham and Fazeley Canal, as well as that canal's link to the Coventry Canal.

During his stewardship as resident engineer of the Trent and Mersey Canal, Thomas Dadford undertook various other building projects and surveys. On 13 September 1788 he wrote to John Sparrow, of the Trent and Mersey Canal, to say he was involved with the rebuilding of Stafford Gaol (although in what capacity is still unknown at present). In the same year he contributed to the cost of building St Peter's chapel, Roman Catholic, at Rushton Grange in Corbridge. This chapel was demolished in 1936.

In 1789 Dadford and Sheasby tendered for the work in constructing the Cromford Canal. They acted in this canal project as contractors working for William Jessop (1754–1814), consultant engineer, and Benjamin Outram (1764–1805), resident engineer. The Cromford Canal Company had offered the work in three contracts: a section from Cromford east including two aqueducts and deep cuttings, this to be completed in two years; Butterley Tunnel and the approach cuttings which constituted the central section, to be completed in three years; and finally a link from the tunnel to the east to join the Erewash Canal and thereby, via the Derby Canal, to the rest of Brindley's system. This final section was to be completed in eighteen months.

Dadford and Sheasby won the contract for building the whole canal but had badly underestimated the magnitude and cost of the project, particularly as they had started work in 1790 as contractors to the Glamorganshire Canal nearly 200 miles away. Like Brindley and Whitworth, Dadford was now travelling somewhere in the region of 10,000 miles a year in all weathers on hazardous roads. By 1790 Dadford and Sheasby were in financial difficulties in Derbyshire. As the Cromford Canal Company held the contractors to the terms of the cost of construction as agreed, Dadford and Sheasby declared it was impossible for them to proceed without a further release of funds. When monies were not forthcoming, Dadford and Sheasby walked off site.

There was no clause in the agreement with the Cromford Canal Company insisting on the satisfactory completion of the work by those contracted to the building. Skempton and Hadfield record that Dadford and Sheasby were overpaid by £1,322, and that the money seems not to have been returned. Schofield, in his book on Benjamin Outram, comments that it was to their discredit that the contractors left. Dadford and Sheasby went off to work on the Glamorganshire Canal, and Benjamin Outram completed the Cromford Canal under Jessop's direction.

Following the debacle of the Cromford Canal, from which, surprisingly, Thomas Dadford appears to have escaped from with his reputation intact, Dadford concentrated on work which involved his sons and Thomas Sheasby on canals and tram roads in South Wales. On 30 June 1790 the Glamorganshire Canal Company met at the Cardiff Arms Inn near Cardiff, and engaged Thomas Dadford and his eldest son Thomas, plus Thomas Sheasby, as joint contractors to construct the canal for £48,288. The canal was managed by a joint committee of ironmasters and landowners without a principal engineer, the role that Dadford actually occupied. This was an unusual arrangement at that time, although Peter Cross Rudkin reminds us that today, design-and-build contracts are not uncommon. Nevertheless the agreement between Dadford, Sheasby and the Glamorganshire Canal Company was to lead to difficulties later in the period of construction.

Richard Crawshay (1739–1810), an ironmaster from Merthyr, led the committee and presumably Thomas Dadford had been recommended to him to by another ironmaster from the Midlands, Francis Homfray. Homfray had an ironworks at Broseley and a forge at Stewponey on the Staffordshire and Worcestershire Canal. He would have met with Thomas Dadford during the production of ironwork necessary for constructing the locks on that canal. Crawshay evidently thought highly of Dadford. John Norris' researches revealed that, in April 1791, Richard Crawshay agreed with Homfray's recommendation when he wrote to Count de Reden in Berlin of his surveyor and 'Navigation Maker' in glowing terms:

> On my return from Wales I found your favour of 26th Febry. – when your engineer appears here I will put him under the Tuition of the ablest Man who has promised me he will teach him all he knows I mean Dadford our Contractor, he has made so great progress that had I not lately been an Eye Witness of any report would have fall'n far short of …

His enthusiasm would be tempered by events that were to follow, but it is interesting to note that the method of tuition by one engineer visiting a site to view and discuss engineering techniques was developing. James Brindley would have been gratified to note his training methods were being taken forward.

The Glamorganshire Canal was the first major canal in Wales to be built, and it brought the coal and iron from Merthyr Tydfil, near Cyfarthfa, to Cardiff. Despite his title as contractor it was Dadford's first canal on which he was exclusively responsible for surveying, engineering and building the line. At the time of their appointment Thomas Dadford and his son Thomas were described as residing at Redland, near Bristol, in the county of Gloucester. It is not known at the time of writing what the Dadfords were doing here.

Dadford was leading a peripatetic life, typical of a canal engineer during the 'Canal Mania', but he seems to have collected properties in Wales and the Midlands. He and Thomas Jnr also each bought £500 of shares in the Glamorganshire Canal. This investment was to reward the two engineers handsomely. At Pentrebach, encouraged by Crawshay, Thomas Dadford Jnr and his wife Ann built an inn. (Crawshay wanted to stay there on his periodic visits to Wales from London.) The inn was named The Duke of Bridgewater and it is more than likely that Thomas Dadford Snr also resided there during the canal's construction. Sadly, it no longer exists.

Navigation House, Abercynon, Glamorganshire Canal, built by Thomas Dadford.

Thomas Dadford arrived with his workmen on the site of the canal in August 1790. He and Thomas Sheasby then began construction in the north at Merthyr Tydfil and worked southwards towards the coast. The canal was 24½ miles long, falling by 543ft to Cardiff through fifty locks, most of them well over 10ft deep, taking boats 60ft by 9ft which carried up to 25 tons each. By 1794 the canal was open to Cardiff. Like all the canals in South Wales it was built down the steep valleys requiring long flights of locks; Dadford's speciality. Nantgarw flight was a prime example of triple locks on the Glamorganshire Canal. There were few tunnels but rivers had to be bridged and the Dadfords constructed several aqueducts. The canal was a fine example of dextrous surveying and imaginative construction, although there were problems with the water supply. This was due to the canal's success and the round-the-clock usage, seven days a week. Rowson and Wright have commented that, strangely, the canal is still a contour canal of the Brindley variety; hugging the sides of the valley before the large drops in level were catered for by flights of deep locks in pairs and sometimes trebles.

Thomas Dadford and Sheasby agreed with the Glamorganshire Canal Committee to construct the canal for £48,288. The contract insisted Dadford and Sheasby give surety of their work by giving the company a bond of £10,000 – whether rumours had spread from Cromford is unknown. The success of the canal's construction and its subsequent trading ability was later marred by a serious rift between Dadford, Sheasby and the canal's management committee. During the construction period, Dadford and Sheasby

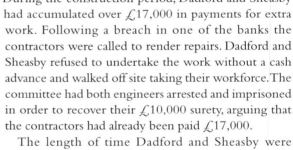

had accumulated over £17,000 in payments for extra work. Following a breach in one of the banks the contractors were called to render repairs. Dadford and Sheasby refused to undertake the work without a cash advance and walked off site taking their workforce. The committee had both engineers arrested and imprisoned in order to recover their £10,000 surety, arguing that the contractors had already been paid £17,000.

The length of time Dadford and Sheasby were incarcerated is not known, but gaols in eighteenth-century England were filthy, graffiti-ridden places where unruly prisoners were herded together with no privacy. There was little access to fresh water and prisoners had to pay the gaoler for their own food and for any other service rendered. In many gaols, cells were divided between debtors' cells and those who were condemned to death. In the eighteenth century over 200 crimes were punishable by death, including the crimes of picking pockets and cutting down trees. The League of Penal Reform was over half a century away, and Dadford and Sheasby must have endured an appalling time.

Chapel Row, Merthyr Tydfil. 'Tombstones', canal distance markers, are preserved here where the Glamorganshire Canal started.

The complex argument over the surety was brought to arbitration. Robert Whitworth, the eminent engineer from the Brindley school who was justly reputed for his integrity as well as his engineering skill, found, in his analysis of events, that Dadford and Sheasby were liable for only £1,512 of the £17,000 claimed. However, Dadford and Sheasby had lost their jobs and other work during their period of imprisonment, and the final section of the canal was completed by Patrick Copeland. By 1798 the sea locks had been built and the company could exchange cargoes with coastal vessels. The argument over payment to Dadford and Sheasby must also have affected the business of Dadford's two sons, James and John, who had become carriers on the canal. They left and returned to the engineering of other Welsh canals and tram roads.

Little is left of this highly successful canal today with the exception of Navigation House in Abercynon. This was the headquarters of the canal built by Dadford Snr in 1792 and it was from where he conducted operations. At Abercynon, the Aberdare Canal and the Penydarren tram road met the Glamorganshire Canal. In 1804 Richard Trevithick's first locomotive ran on this tram road and its presence signed an early death warrant of canal construction in the country; heralding as it did the arrival of the railways in just over thirty years' time following the end of the Napoleonic Wars.

Sadly, the great Nantgarw treble locks engineered by Dadford and his son have been obliterated by the construction of a new road. Each of these locks was 14ft deep and they were illuminated by gaslights for round-the-clock working during the canal's busiest period. At Aberfan Dadford created the deepest lock in Britain, greater by 6in than the top lock of the Tardebigge flight on the Worcester and Birmingham Canal. John Bird, writing in 1796, said: 'The Canal is brought through mountainous scenery with wonderful ingenuity.'

Thomas Dadford Snr survived the acute unpleasantness of his confrontation with the Glamorganshire Canal Company and moved further north in Wales, taking over

Tramway wagon from the Glamorganshire Canal Company at Merthyr Tydfil, preserved at Chapel Row.

Montgomeryshire Canal below Carreghofa Locks.

the engineering of part of the Montgomeryshire Canal from his sons John and Thomas after the former had immigrated to America in 1796. The Dadfords completed the section from the junction with the Ellesmere Canal at Carreghofa to Garthmyl, south of Welshpool. Two aqueducts had been built at Vyrnwy and Rhiw which caused problems in construction. Following a partial collapse of the Vyrnwy Aqueduct, John Dadford, who had been blamed in part for the collapse, emigrated to America with Thomas Wedgwood.

On Thomas Dadford Jnr's recommendation, old Thomas Dadford 'under whom they were bred and who has great experience of canal works' was called to complete a section of the Montgomeryshire Canal, including the Vyrnwy Aqueduct. Whether this was a form of restitution to the world of engineering for Thomas Dadford Snr by his sons, or whether with the departure of John the company urgently needed a replacement engineer with sound practical knowledge, is not known. The latter seems most likely. Lime and limestone were the main traffic on the canal; the limestone being burnt in kilns by the canal and then distributed via the canal for agricultural purposes. The aqueduct at Vyrnwy continued to cause trouble and eventually George Watson Buck repaired it using wrought-iron tie rods and cast-iron facing plates together with cast-iron beams on the actual arch face

The Dadford family built many of the Welsh canals. Dadford's sons had worked on canals in South Wales and in the border country. Thomas Dadford Jnr was responsible for the Neath Canal, the Monmouthshire Canal with its great flight of locks at Rogerstone on its branch to Crumlin, and the Brecknock and Abergavenny Canal, as well surveying the Aberdare Canal. He also engineered the Leominster Canal, which was never completed. This canal was designed to join Kington to the Severn at Stourport, thereby opening Leominster to manufactured products from the Birmingham area.

Vyrnwy Aqueduct,
Montgomeryshire Canal.

St Teilo's church, Llanarth. The
graveyard contains the tomb of
Thomas Dadford Jnr, Thomas
Dadford's eldest son.

Thomas Dadford Jnr died at the age of forty on 2 April 1801 at Crickhowell, as 'a result of a violent fever upon the brain brought on it is supposed by a cold which he had taken during that inclement season'. He was buried at St Teilo's church, Llanarth, in Monmouthshire and his tomb may be visited today. James Dadford, the second eldest son, having worked on the Glamorganshire Canal building boats for the transportation of coal and iron engineered the tram road from Aberdare via Hirwaun to Penderyn, before returning to the Midlands. It is thought he worked on the Staffordshire and Worcestershire Canal. Before he left Wales he had also been employed to make good his older brother Thomas' engagement with the Brecknock and Abergavenny Canal Company after the latter's death. On 20 February 1804 James died at Stourport 'after a long and painful illness', his death 'lamented by a large circle of friends and acquaintance'. This was a tragedy for Thomas and Frances in that at least two of their sons had predeceased them and the other two had left for America; one definitely in search of work, possibly constructing canals, and the other probably to find his older brother.

By 1799 Thomas Dadford had undertaken his last engineering work in Wales, surveying a slate railway at Penryn in the north of the principality. Dadford Snr was now on his own; his sons either dead or departed to the New World. Seven years earlier, in 1792, Robert Thompson had written to William Tait to report meeting Thomas and James Dadford. He wrote, 'Old Dadford shuffled without saying any thing to the purpose', and his son James said 'not to trouble his father'. John Norris says of this meeting that it could be an indication of Dadford getting old. He was; but it was also likely that Dadford did not want to get involved with aiding Tait out of loyalty to Crawshay, Dadford's patron and Tait's rival. Dadford certainly played his business cards close to his chest throughout his career and was unlikely to prejudice anything which might cost him money or advancement.

Norris also postulates that the younger Dadfords may have suffered from tuberculosis, which seems very possible. Increasing age and the opportunity to be near James Dadford (before his death) in Stourport, and his daughter Mary in Wolverhampton, meant Thomas Dadford Snr returned to the Midlands. In 1800–01 we hear of him designing a chapel for the Roman Catholic school at Sedgley Park where his three youngest sons had been educated. The chapel still exists and is now an Indian restaurant, part of the Ramada Hotel at Sedgley Park.

Thomas Dadford Junior's grave, St Teilo's church, Llanarth.

Thomas Dadford was now an old man who had accrued considerable wealth, held as properties, shares and land as a result of his extensive engineering and investment activities. On 14 June 1809 Dadford suffered a further blow when his wife Frances died. They had been married for almost fifty years and were a wealthy couple in financial terms. The wealth was largely held in lands, tenements and bonds. On 13 September 1809 with 'times winged chariot hurrying near', Thomas Dadford made a new will. His executors, Anthony Lane and James Marsh of Wolverhampton, were instructed to sell his properties and lands and use the proceeds to buy Government Securities which would yield half-yearly dividends for his beneficiaries after his death.

For his daughter-in-law Ann Dadford, the wife of Thomas, these securities were authorised to yield £296 14s 6d in interest 'to be paid to her for the term of her natural life by half-yearly payments'; a generous allowance. The same arrangement was to be put in place for his 'dear daughter Mary [Rock]', and a trust was created to pay monies to William Dadford or his descendants who may have been in America. If, after seven

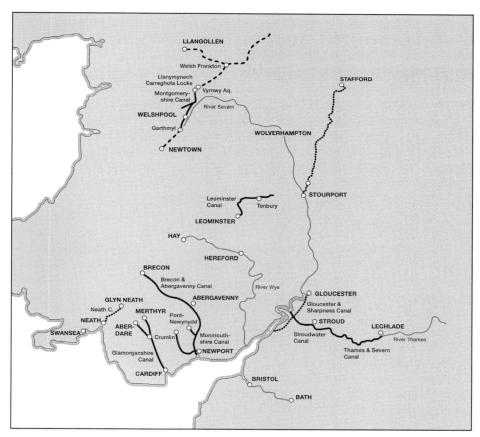

Fig.2: Dadford's canals in Wales.

	Thomas Dadford and his sons' canals in Wales
	Canals partly built by the Dadfords' in Wales
	Canals constructed by other engineers

Sedgley Park School chapel, designed by Thomas Dadford Senior, 1800–01.

years, he or they had not returned to England then the estate became Mary's. Interestingly, no share of the estate was left to John Dadford. Had there been a rift between father and son over the Montgomeryshire Canal works and the emigration to America? Or had news reached Dadford from his youngest son William that John had died, without this being recorded? Thomas Dadford seemed uncertain as to the whereabouts of his sons in America.

John Dadford and Thomas Wedgwood from Burslem were sharing a log cabin in Pennsylvania until March 1798 when they became discouraged and left. The *Staffordshire Advertiser* of 16 March 1799 recorded the death of Wedgwood in New York in September 1798 as a result of yellow fever. Given the health record of the Dadford sons, did John also succumb? Did Thomas Dadford Snr know this?

In October 1809 Thomas Dadford Snr died at the remarkable age of 79, and was buried at St Peter's church, Wolverhampton on 28 October 1809. He had outlived two, possibly three of his sons and his wife Frances. It is not known if Thomas was buried with Frances – who he had survived by only four months. Their tomb or tombs no longer exist at St Peter's church, although the church itself and the centre of Wolverhampton are well worth a visit.

Like our other engineers we have no portrait of any of the family and we know nothing of Thomas Dadford Snr's early life at present. Norris' verdict is that:

> Dadford was a significant engineer of the first period of canal building in Britain. As well as his work on the canals of the Black Country and the Grand Cross, the canals and associated tramroads constructed by the Dadfords and Sheasbys down the Welsh valleys to Swansea, Neath, Cardiff and Newport made a major contribution to the industrial growth in South Wales.

Without Thomas Dadford and his sons' contributions, the canals in the Midlands and South Wales would not have been built with the speed, industry and overall competence that they were. These canals furthered greatly Britain's industrial development and economy. It is unfortunate that Thomas Dadford's finest piece of canal surveying and engineering at Glamorgan should have been prejudiced by bitter controversy over contracts and obliterated by time. However, thanks to John Norris, we know much more of the lives of the Dadfords and, it is hoped, more will emerge in the future.

7

John Varley, Samuel Weston and James Brindley Jnr

'Amongst all the heroes and all the statesmen that have ever yet existed none have
accomplished anything of such vast importance to the world in general as have been
realized by a few simple mechanics.'
– James Sims, 1849

John Varley 1740–1809

John Varley, surveyor and engineer, was an important assistant to James Brindley in setting out his canals in the East Midlands that were linked to the Grand Cross, and later, after Brindley's death, the Leicestershire and Northamptonshire Union Canal with William Jessop. He can reasonably be declared as a member of the school of engineering given his selection and training by Brindley when working on the Chesterfield Canal.

Varley was born in Heanor in Derbyshire in 1740 and was trained as a surveyor. Christine Richardson thinks that Varley attended two 'workcamps' which were based on Brindley's two leading projects, the Bridgewater and the Grand Trunk canals, early in his career. These camps on working projects were to train and set out principles for the building of canals in the manner Brindley approved. It is more than likely his training under Brindley's watchful eye took place near Runcorn on the Bridgewater Canal because he married Hannah Patters at Dutton there.

Brindley was quickly made aware of Varley's abilities when he used him to assist with a survey of the Chester Canal. Although this canal was authorised to be built on a different line in 1772, in 1769 John Varley surveyed the Chesterfield Canal with James Brindley. As a result of this successful work Brindley sent him to replace Robert Whitworth on a survey of the Leeds and Liverpool Canal in 1769.

Varley surveyed ten canals, two for Brindley, and was the resident engineer of the Chesterfield Canal during Brindley's lifetime. After Brindley died he worked for Hugh Henshall – although there were difficulties during the changeover of consulting engineers regarding some members of the Varley family. This included John's father, Francis, and a brother, who were sacked from the Chesterfield Canal works for dishonest claims for labour and shoddy workmanship in the Norwood Tunnel. John Varley was lucky to keep his own job in these circumstances, but expert canal engineers were in much demand and short supply.

THE VARLEY FAMILY

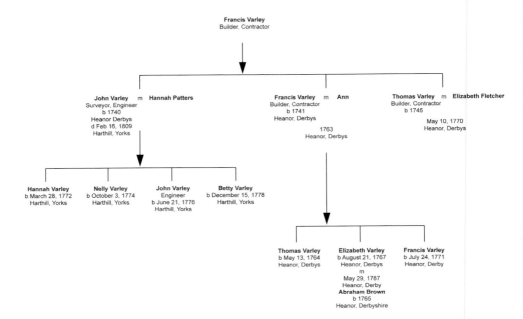

Varley completed the Chesterfield Canal, which was 45 miles long, with fifty-nine narrow locks and six wide locks. The Thorpe flight had fifteen locks and the Norwood flight thirteen; several of the locks were staircase locks. He also engineered the great summit tunnel at Norwood which was the longest of its period being 2,884yds in length. Mike Chrimes records that the work on the Chesterfield Canal was done on a low budget, with minimal help from Hugh Henshall, who was busy completing the Grand Trunk Canal and Brindley's other unfinished works. However we know Henshall organised efficient working practice on this canal.

In 1774 John and Hannah Varley had a house built by the canal company at Pennyholme near the tunnel: the remains of the foundations can still be traced. Both lived in the house until their death and raised six children there, although after the completion of the Chesterfield Canal Varley was involved with many other engineering projects. In particular Varley went on to engineer and survey the Erewash Canal, the Nutbrook Canal, the Nottingham and Beeston Canal and the Leicestershire and Northamptonshire Union Canal. The latter, worked on with Christopher Staveley, was 40 miles long and required four tunnels and thirty-eight locks. Varley was also the contractor to build Wolseley Bridge over the Trent near Colwich, where the original plans for the building of the Grand Trunk had been agreed in 1767. He did not finish this work which was completed by direct labour. We last hear of him subcontracting to undertake repair work on the Huddersfield Narrow Canal in 1800–01. He was supplying mortar and subcontracting work to complete the great summit tunnel at Standedge.

Christine Richardson, in her excellent book on the Chesterfield Canal, *The Waterways Revolution*, reveals that by 1808 Varley was 'in a very infirm and distressed state and

Chesterfield Canal, near Worksop, during reconstruction.

soliciting an Annuity from the Erewash Canal Company'. The annuity was refused but a one-off donation of 20 guineas was made. Varley lived at Pennyholme, Kiveton Park, from 1773 until his death in 1809. He was buried at Harthill in South Yorkshire on 16 February 1809, close to his first canal tunnel at Norwood.

Samuel Weston *c.*1730s–*c.*1804

Samuel Weston, surveyor, engineer and contractor, probably came from Cheshire, possibly near Runcorn. Mike Chrimes reveals he worked for Brindley on one of the early 'workcamps' on the Bridgewater Canal as a staff holder and surveyor. On 2 October 1757 he married Mary Ankers at Aston-by-Sutton church.

Weston also surveyed and levelled for James Brindley on the canal line from Chester to the Trent and Mersey Canal at Middlewich. The canal was not built, however, due to lack of parliamentary approval. A second attempt to build the canal also failed due to lack of parliamentary assent in 1770, but by 1772 an amended route, that did not connect with the Grand Trunk Canal, had been authorised. Weston worked on this project until 1774 and was replaced by Thomas Morris, who in turn was replaced by Josiah Clowes. Weston's leaving was possibly related to the partial collapse of the Gowy Aqueduct. Before Parliament, Weston disarmingly admitted he had never built an aqueduct before.

Whilst working for the Chester Canal Company Weston was also employed on the Leeds and Liverpool Canal as a contractor for 4 miles of construction between Newburgh and Liverpool. He also joined with Samuel Simcock, whom he may well have met during his training period on Brindley's early canals, to build the Oxford Canal. He was involved with the proposal to make the Cherwell navigable to Oxford, and later the new line from Banbury to Oxford that was built. He also assisted Simcock with the survey of the Hampton Gay Canal, a cut from Thrupp near Oxford to Brentford, to obviate the vicissitudes present in navigating the Thames. He reported to Parliament with Simcock and Henshall regarding the projected canal but in the event the navigation was never built.

In 1788 Weston was asked to survey the Western Canal with Simcock and Barnes. This canal was the forerunner to the Kennet and Avon Canal and was planned to link the two rivers in a single navigation. The Great Western Canal route they proposed was a more northerly route than the one actually built by John Rennie. Robert Whitworth, who was also consulted, drew attention to the lack of water supplies at the summit level of the proposed Weston/Simcock/Barnes canal. This situation was later to bedevil Rennie's canal as built.

In the last decade of the eighteenth century Weston had set up business as a contractor with his son William (1763–1833) in Oxford: 'Samuel Weston and Son.' They successfully tendered to build part of the Ellesmere Canal from Ellesmere Port to Chester, plus the Llanymynech branch to connect the Montgomeryshire Canal. Samuel Weston did not complete this branch which was finished by John Fletcher, a partner. Weston died between October 1804 and March 1805.

However William Weston, his son, went to America in November 1792 and engineered canals, bridges and turnpike roads there. He must have met James Brindley Jnr because he worked on a contract to the Schuylkill and Susquehanna Navigation in Pennsylvania. He, like Brindley Jnr, met George Washington who requested that he construct a link between the Hudson River and the Great Lakes in upstate New York. This was the precursor of the Erie Canal.

William Weston left America after nine years and declined to take on the building of the Erie Canal, although he continued to give consulting advice to that canal company. He settled in Gainsborough on his return but died suddenly in London on 29 August 1833 of 'ossification of the heart'.

THE WESTON FAMILY

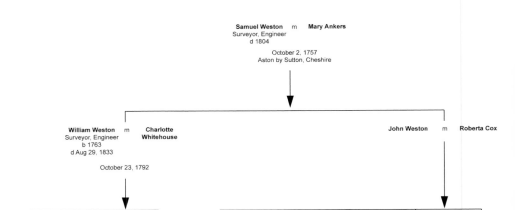

James Brindley Jnr 1745–1820

James Brindley Jnr, surveyor and engineer, was born in 1745 – a year before his parents were married. He was baptised on 8 May 1748 at Waterfall in Staffordshire. His parents were James Brindley's younger brother, Joseph Brindley, and Sarah Bennett, who married in Alton on 17 December 1746. Joseph Brindley worked as a millwright and was reported to be working in a smelting factory producing wire in Alton in the 1740s.

This branch of the Brindley family lived for a while at Lowe Farm in Leek, presumably with Joseph's parents James and Susannah Brindley. Joseph and Sarah and family at one point also lived at Rough Hey Farm at Gawsworth near Macclesfield in Cheshire.

James Brindley Jnr had one brother, Mathew, and four sisters. Between 1748 and 1762 the family were living at Checkley. Sarah Brindley *née* Bennett had died whilst still young, possibly from repeated childbirth, and Joseph married Lydia Lightwood on 13 February 1762. They had two more children, Henry and Lydia, before Lydia Snr died, again probably in childbirth, and was buried in Leek on 11 November 1763. Joseph then married fifty-three-year-old Mary Mobberley in Leek on 3 July 1764. Romance aside, this move was very possibly in order for Joseph to acquire help with the upbringing of his children by his first two wives. Joseph Brindley died in December 1790 and was buried at Alton on 26 December the same year.

James Brindley Jnr attended Overton Bank, the Quaker school in Leek, which was within walking distance of Lowe Hill Farm. Kapsch and Young, in their detailed *Newcomen Society Transactions* paper on James Brindley Jnr, think it probable that he was apprenticed to his uncle around the age of fourteen as this was typical of the time. If this is true then the younger Brindley could have worked on the Bridgewater Canal with his two uncles, Samuel Simcock and James Brindley, as well as working on the Grand Trunk Canal and its associated canals. He would have been a true pupil in this work and not part of the school of engineering described previously. No mention is made of James Brindley Jnr in canal records in Britain so the work must have been purely tutorial in nature. James Brindley Jnr did claim to have worked on the Bridgewater Canal himself.

On 8 May 1772 John Ballendine, a colonial iron manufacturer, travelled to Europe, encouraged by George Washington and the Governors of Maryland and Virginia, to try to recruit English engineers to come to America to introduce canal and other technologies they had developed in England. George Washington, apart from being a major plantation owner and future first president of the United States, was a surveyor himself and had made a number of intrepid journeys on the Potomac River with a view to making it navigable. Ballendine arrived in London in January 1773 and visited the Bridgewater Canal. A year later, with approximately forty other 'engenious Mechanicks' from Britain and France, James Brindley Jnr was recruited to the colonies and arrived in Hampton on 7 July 1774. With him was his cousin Thomas Allen, also an engineer and the son of James and Joseph Brindley's sister Ann, who had married William Allen, an inn keeper of Leek.

George Washington wrote:

Mr Brindley, nephew to the celebrated person of that name who conducted the work of the Duke of Bridgewater and planned many others in England, possesses, I presume, more practical knowledge of Cuts and Locks for the improvement of inland navigation, than any man among us, as he was executive officer (he says) many years under his uncle in this business.

James Brindley became one of the engineers principally responsible for transferring the technology of canal building to America. He was involved in the building of eight canals there, and the planning of a further two canals that were never built. All the canals were of a short length, normally bypassing rapids or falls on river navigations. The longest canal was the Santee Canal in South Carolina which was 22 miles and linked the Santee and Cooper Rivers. Brindley was chiefly responsible for the survey of this canal which was the first summit canal constructed in America. It was completed by John Christian Senf.

The other cuts Brindley built, apart from the Conewago and Susquehanna canals, were 3 miles or less in length. The first canal was constructed between 1775–76 was at Little Falls on the Potomac River, 3 miles above Georgetown – later to become part of Washington DC. In 1792 James Brindley engineered the Conewago Canal on the Susquehanna River adjacent the to Conewago Falls. It was the first operational canal in Pennsylvania and included what the Americans referred to as 'Lift Locks'. These were two locks, 80ft long and 12ft wide, to accommodate the descent of the Susquehanna River through the falls. When complete the canal was 22 miles long, 35ft wide and 5.6ft deep. With the French Indian War over, entrepreneurs were keen to open out the country for trade in timber, whisky and other commodities.

With his connections to George Washington, James Brindley Jnr became sympathetic to the Colonists' cause. He subscribed to the Oath of Allegiance and Fidelity and joined Captain John Garrett's militia in 1778. A year later, during the Revolutionary War, aged thirty-four, he married Elizabeth Ogle, five years his senior, at Old Swedes church in Wilmington. Elizabeth was well practised at marriage having already had two previous husbands. James, meanwhile, was practised in fathering children; he had a daughter, Catherine Smidt, who was born on 8 November 1779 just seven months after his marriage. The mother was not his wife but Rebecca Smidt.

James and Elizabeth then had three children: Sarah, Susanna and James Joseph (the Brindley family names are evident). They lived in Wilmington, Delaware until James died, aged seventy-two, on 24 November 1820. A large obelisk commemorates the family there.

Conclusion

'The profession of a civil engineer, being the art of directing the great source of
power in Nature for the use and convenience of man'
– Thomas Tredgold, 1828

The great age of civil engineering began in Britain principally as a result of the building of the canals. In the Introduction, mention was made of the two principal engineers at the outset of the canal age. The first was Smeaton. The second was James Brindley, accompanied by members of his 'School of Engineering'. Of the two maps, Figure 3 shows the number of canals constructed at the time of Brindley's death in 1772. Figure 4 shows just how much of the canal system was planned and engineered by James Brindley and his school by the end of the eighteenth century. Of course, members of Brindley's school were also part of other canal work outside Brindley's brief, and several, such as Whitworth, Henshall and Dadford, became members of the Smeatonian Society – indeed Brindley worked with Smeaton himself.

Brindley never joined the Smeatonian Society though. This may have been because he was nagged by a sense of inadequate education, or because he simply did not have time given the ever increasing demands for his consultancy in the building of canals. Although one must share Darwin's admiration of all Brindley achieved in his lifetime, and from such unpropitious beginnings, one becomes slightly suspicious of Brindley's motivation on occasion, particularly when reflecting on his comment to Louis XV of France. Having received an invitation to view the canal of Languedoc, Brindley responded saying: 'He would make no journeys to other countries, unless it were to be employed surpassing what was already done in them.' One is tempted to reply that he would have learned a great deal, particularly about water supply and the creation of reservoirs, as well as the long-term economies that could be achieved by larger scale navigations. However, at least he realised the importance of standardising the gauge.

The methods of building canals were clearly improved by the work of William Jessop, John Rennie and Thomas Telford. Amongst other advances, they chose not to adopt the principal of contour engineering in their canal construction but perfected cut and fill techniques (although it should be remembered that Robert Whitworth was also engaged in this procedure – witness the Burnley Embankment on the Leeds and Liverpool Canal). The new techniques ultimately heralded the civil engineering required in the railway age. In addition, Telford and Jessop introduced, with Benjamin Outram, the use of

Fig.3: canals built at the time of Brindley's death.

iron in the structural engineering of bridges and aqueducts. One of the earliest examples being the Longdon-on-Tern Aqueduct, built by Telford on the Shrewsbury Canal. This structure replaced Josiah Clowes' conventional masonry aqueduct. Later came the awe-inspiring Pontcysyllte Aqueduct built by Telford and Jessop over the River Dee.

James Brindley and members of his school left the country with a very considerable system of canals and a knowledge of canal and civil engineering that influenced all those who followed; paving the way for the surveying and building of the railways that were

Fig.4: canals built by Brindley's school at the beginning of the nineteenth century.

to come. Like any good leader, Brindley laid down a method, albeit partly informal, of apprenticing engineers through his 'workcamps'. This enabled his work to be completed and extended. As we have seen, the men described in this book went on to develop the canal system in this country further! By the end of the eighteenth century over 1,100 miles had been engineered by members of the school and it would not be an unreasonable guess that twice that mileage had been surveyed, particularly if the work on river improvement is taken into account.

Close-up of James
Brindley's statue at Coventry
Basin. (Kind permission of Danny
Hayton)

There were of course other engineers not mentioned in this book. Samuel Bull (active 1768–98) became an associate member of the school of engineering when he was appointed to work on the Coventry Canal as assistant clerk of works. He was sent to the Staffordshire and Worcestershire Canal to gain experience in building locks, exemplifying the Brindley training principles. Later Bull worked on the Oxford Canal but ended his career working principally for the Birmingham Canal Company on a wide range of engineering tasks. These canals were all Brindley canals. In the 1780s Bull engineered the junction between the Birmingham Canal and the Dudley Canal, near Tipton.

The British approach, espoused by James Brindley and individual companies, stood out from the French school of canal building. Michael Chrimes has observed that 'Sully and Colbert had seen advantage of a co-ordinated communications network as a basis for commercial prosperity'. They had promoted investment in the building of French canals and improving navigable waterways. Yet 'by 1800 Great Britain had a tremendous lead in engineering over the rest of Europe'.

The competiveness and ingenuity of James Brindley and his school of engineers, and the ambitions of the canal companies and associated industries, went a long way to fulfil Thomas Tredgold's 1828 description: 'The profession of a civil engineer, being the art of directing the great source of power in Nature for the use and convenience of man.' This sums up well the role of Brindley's school of engineering in providing Britain with a vital new transport system.

Gazetteer

This gazetteer is an introductory guide. Development, change and preservation can alter what is to be seen on the ground quickly. It is recommended to use the information here as a start, and plan visits accordingly.

James Brindley

Works 1760–72

1760–63	Bridgewater I	Surveyor, Resident Engineer, Chief Engineer
1762–72	Bridgewater II	Surveyor, Chief Engineer
1766–72	Trent and Mersey Canal	Surveyor, Resident Engineer, Chief Engineer
1766–72	Staffordshire and Worcestershire Canal	Surveyor, Chief Engineer
1768–71	Droitwich Canal	Surveyor, Chief Engineer
1768–72	Coventry Canal	Surveyor, Chief Engineer
1768–72	Birmingham Canal	Surveyor, Chief Engineer
1769–72	Oxford Canal	Surveyor, Chief Engineer
1771–72	Chesterfield Canal	Surveyor, Chief Engineer

Life 1716–72

All the references given below are taken from the Ordnance Survey Landranger maps, 1:50,000 (2cm:1km)

Brindley's Croft. OS Sheet 119 Brindley's birthplace is at Great Rocks, Tunstead, in Derbyshire, 3 miles north-east of Buxton. The cottage is completely derelict but a memorial stone can be found amongst the ruins commemorating Brindley's birth and life achievements. Also, at Wormhill, there is a well built in memory of James Brindley, **Ref 114742**. A plaque records he was born in the parish in 1716.

Brindley's Mill. OS Sheet 118 Leek, Staffordshire. This watermill has been restored as a museum and is where Brindley worked for a time before going on to found the canal system in England. The mill dates from 1752 and is powered by two undershot wheels, 16ft in diameter. The mill stands on three floors and is the only preserved watermill with associations with James Brindley that survives. It lies ¼ mile north-west of Leek on the A523 road, **Ref 978570**. Enquiries before visiting are recommended as it is not always open.

Road arch, originally through the Barton embankment, Bridgewater Canal.

Bridgewater Canal 1. OS Sheet 109 The whole of Brindley's first canal can be walked from the mines at Worsley to the Castlefield basin in Manchester. Barton Aqueduct has been replaced by a swing bridge, built by William Leader Williams, to carry the Bridgewater Canal over the Manchester Ship Canal. One of the archways from the original aqueduct that spanned the road still exists and forms part of the embankment to the present Bridgewater Canal. There is also a support wall remaining on the northern embankment for the original Barton Aqueduct; this was reconditioned in the nineteenth century.

Bridgewater Canal 2, Bollin Embankment. OS Sheet 109 This magnificent embankment on the second Bridgewater Canal can be visited to the west of Tatton Park, **Ref 728875**. The dimensions of the build recall the height and width of Shelmore Great Bank, built by Telford at the end of the canal age. The embankment is approximately 35ft high and it makes an interesting comparison with the embankments and scale of canals built by Brindley on the Great Cross after the Bridgewater Canal had been completed. Where the embankment crosses the River Bollin, there is what is best described as a culvert enabling the river to pass under the structure. The arch of the culvert has been added to by a hugh battered wall and supporting bridge. A good walk from the modern aqueduct along the embankment and then back via Tatton Park may be had.

Droitwich Canal. OS Sheet 150 Brindley surveyed this early broad canal with Robert Whitworth, built by John Priddey which opened in 1771. This barge canal carried salt from Droitwich as well as coal to Droitwich. The canal reopened in July 2011 with the Droitwich Junction Canal. The latter linked the Brindley canal to the Worcester and Birmingham Canal. Much of the 6¾ miles of the canal can be walked. One of the most dramatic views of this canal is obtainable from the bridge over the deep cutting in Salwarpe by the church, **Ref 875620**.

Wolseley Arms. OS Sheet 128 Public house, still in existence, where the meeting took place to discuss the proposals for the building of the Grand Trunk or Trent and Mersey Canal. James Brindley spoke at the behest of Josiah Wedgwood. Lord Gower, local MPs and other interested parties were present. Hugh Henshall was presumably in attendance as he had drafted the map of the proposed waterway. Nearby is Shugborough Hall, seat of Lord Anson, supporter of the venture. A good description of the meeting is recorded in J. Phillips' *History of English Navigation*. **Ref 201022.**

Harecastle Tunnel. OS Sheet 118 Completed by Hugh Henshall after Brindley's death, the tunnel took a further four years of construction before it was navigable in 1777. The tunnel is 2,880yds long, 9ft wide and 12ft high. It was the first substantive tunnel to be built in this country for transport purposes; previous tunnels wound in seams in pursuit of minerals. Additional branch canals of smaller gauge were built into the hillside to retrieve coal in the manner practised at Worsley Delph on the first Bridgewater Canal. Today, canal boats go under Harecastle Hill in Telford's new tunnel, but both the north portal, **Ref 842542**, and south portal, **Ref 851518**, of the original Brindley/Henshall Harecastle Tunnel can be seen, and there is a fine walk over the top of the tunnel via the old towpath. The ground above at Golden Hill was owned by the quintet of engineers and businessmen Hugh Henshall, James and John Brindley and John and Thomas Gilbert.

Turnhurst Hall. OS Sheet 118 Home of James Brindley and his wife Anne from 1768. Josiah Wedgwood was a regular visitor there. Sadly, the hall no longer exists but there is a public house dedicated to James Brindley at Turnhurst. There are also excellent photographs of the hall and the summerhouse where Brindley worked. These can be found in Klemperer and Sillitoe's account of their archaeological excavation to discover whether Brindley had built a model canal and lock in the grounds of the hall, **Ref 865530**. It is reputed that the door of the hall was removed to Chell primary school and survives.

Dove Aqueduct. OS Sheet 128 East of Burton-on-Trent Dove Aqueduct is to be found, built by James Brindley and Hugh Henshall. It possesses twelve arches with 15ft 4in span. Built between 1768–70, it can be found by a walk along the towpath from the A5121 to the west, **Ref 268269**.

Red Lion Inn, Ipstones. OS Sheet 119 The inn where Brindley was placed in a damp bed during his survey of the Caldon Canal still exists as a friendly public house and hotel in this small village, **Ref 019499**. The beds are now well aired. Unfortunately, there is no plaque to commemorate the great engineer's stay, although that might not be considered good for business, given the circumstances in September 1772.

Tomb at St James church, Newchapel. OS Sheet 118 At the end of a row of Henshall graves, to the left of the church, the grave of James Brindley can be found next to a hedge; a far cry from Westminster Abbey. The tomb has had a memorial tablet fixed to it which was funded by public subscription. Next to James Brindley lies his wife, buried with her second husband, Robert Williamson, and alongside members of the Henshall family, including Hugh, **Ref 862548**.

Statues of Brindley

There are several statutes of James Brindley. One, from which the cover picture of the book is taken, is to be found at the basin of the Coventry Canal in Coventry. Near Etruria Locks in Stoke there is another, sculptured by Colin Melbourne and unveiled in 1990. Brindleyplace, in Birmingham, commemorates the contribution the great engineer made to the development of the city.

Hugh Henshall

Works 1766–1806

1766–77	Trent and Mersey Canal	Clerk of Works, Chief Engineer, Surveyor in succession to Brindley (1772)
1772–77	Bridgewater Canal, link to Runcorn	Engineer in succession to Brindley
1772–77	Chesterfield Canal	Chief Engineer in succession to Brindley
1776–79	Caldon Canal	Engineer in succession to Brindley
1790–91	Manchester, Bolton and Bury Canal	Chief Engineer, Surveyor
1792	Herefordshire and Gloucestershire Canal, Gloucester–Ledbury	Surveyor
1793	Grand Western Canal	Surveyor with William Jessop
1806	Glamorganshire Canal	Surveyor

Life 1734–1816

*N.B. Another useful guide to various places associated with Hugh Henshall in Stoke and the surrounding five towns is published as an A–Z street atlas: Stoke-on-Trent/Newcastle-under-Lyme. There are at least two areas named after Henshall or his relatives. Henshall Place at Sandyford, south of Golden Hill, **Ref ST6**, where Henshall owned lands and mines over Harecastle Tunnel; and Henshall Road near Crackley, **Ref ST5**, in Newcastle-under-Lyme near to where Henshall went to school. There is also Dr Henshall Hall, **Ref CW12**, near Congleton and the Macclesfield Canal. The connection between Hugh Henshall and Dr Henshall is unknown.*

St Bartholomew's church, Norton-in-the-Moors. OS Sheet 118 Hugh Henshall attended Josiah Clowes' wedding here in 1762 as his best man. Buried in the churchyard are Hugh's older sister Jane, who married coal owner William Clowes, Josiah's older brother; and also Josiah Clowes himself, the engineer who worked regularly with Henshall on canal construction, **Ref 894514**.

Findon: Plum Pudding House. OS Sheet 128 'A favourite watering hole', meeting place for Henshall, Lingard and Samuel Simcock to discuss the production of bricks for the building of the Grand Trunk Canal. Plum Pudding House is on the A513 Road, west of Armitage, where the Trent and Mersey Canal tunnelled originally. The tunnel has now been opened out. Armitage Park, which lies near, was the property and seat of Josiah Spode Esq., the potter. Plum Pudding House is now a fashionable Italian restaurant.

Greenway Bank Farm. OS Ref 118 Hugh Henshall had the farm built and he left it in his will to his niece, James and Anne Brindley's eldest unmarried daughter Anne, for the duration of her lifetime. It is now a country park near Rudyard Reservoir, where Rudyard Kipling was conceived, **Ref 889551**. The stables remain, which presumably were used to provide the horse power for the Henshall canal carrying company which operated on the Trent and Mersey Canal. In 1897 Hugh Henshall Williamson, Hugh Henshall's nephew, died there. He had become the High Sheriff of Staffordshire and possessed a fortune created from mining, pottery and other works in the area.

Henshall Graves, St James, Newchapel. OS Sheet 118 In St James churchyard, next to the tomb where James Brindley lies buried, are four tombs belonging to the Henshall family. To the left are Hugh's parents, John and Anne, in a grave with his grandfather also named Hugh Henshall. Hugh Henshall, the engineer, is referred to as 'being of the firm Hugh Henshall and Company of the Trent Mersey Canel, late of Longport'. (Note the spelling of canal.) Hugh is buried in a tomb with his second brother-in-law, Robert Williamson, and his nephew, John Williamson, who tragically drowned in an accident. His sister, Anne Williamson, was later buried in the same plot with her brother, son and second husband, **Ref 862548**.

Little Ramsdell Hall. OS Sheet 118 Purchased by Hugh's nephew Robert, son of Anne who had married James Brindley and then Robert Williamson. This striking L-shaped red-brick Georgian building overlooks the Macclesfield Canal. It was built between 1720 and 1760. After Robert Williamson died it became the home of John Dobbs, the late eighteenth-century inventor and engineer. It is a private residence but can best be viewed from the footpath of the Macclesfield Canal, **Ref 839582**.

Caldon Canal. OS Sheets 118–119 Hugh Henshall surveyed and engineered this canal following Brindley's death. This beautiful canal is now open to Froghall, and the Uttoxeter Canal which it joined has closed, but a basin and lock have been restored and much of the system can still be walked. The canal, with a branch to Leek fed by the Rudyard Reservoir, was built to bring limestone from the Caldon Low quarries, and flints from Cheddleton to the Potteries at Stoke. It has examples of staircase locks at Bedford Street, the only examples in north Staffordshire, and a flint museum at Cheddleton Mill. At the Bedford Street Locks there is a statue of James Brindley. Hugh Henshall also built tramways to connect with the canal. The website of the Caldon and Uttoxeter Canal Trust is www.cuct.org.uk.

Section of Trent and Mersey Canal between Barnton and Preston Brook. OS Sheets 118–117 The tunnels at Barnton, southern portal, **Ref 638749**, Saltersford Tunnel and the northern section of the Trent and Mersey Canal were entirely re-surveyed and built by Hugh Henshall after Brindley's death. This section of the canal joined the Duke of Bridgewater's canal at Preston Brook. Preston Brook Tunnel was also engineered by Henshall. The section affords an interesting walk with magnificent views from the Trent and Mersey Canal above the River Weaver. The tunnels at Barnton and Saltersford are of particular note as they are the sole surviving examples of Brindley's narrow-gauge tunnels from the Grand Trunk Canal; Harecastle being closed. Walkers can also visit the Anderton Lift, built by William Leader Williams in the nineteenth century. This took craft from the Trent and Mersey Canal down to the River Weaver.

Bent Farm, Newchapel. OS Sheet 118 This former home of the Henshall family is now demolished. However, there are photographs of the farm house, taken by A.R. Saul in the 1920s, that can be viewed at the William Salt Museum in Stafford.

Etruria Hall. OS Sheet 118 This was the place on which Wedgwood built his original factory for the production of pottery and porcelain. The Trent and Mersey Canal was laid out by Henshall, after Brindley's survey, in front of the works. Both Wedgwood and Henshall adhered rigidly to the plans as any deviation would cause an eruption of temper on Mr Brindley's part. Photographs exist of these majestic works with the canal in front, although the works were demolished in 1968. Wedgwood's factory is now situated to the east at Barlaston and is well worth a visit.

Longport Pottery Works. OS Sheet 118 Henshall's, Williamson's and Clowes' pottery works were based at Longport. The warehouse buildings are still extant by the side of the Trent and Mersey Canal, approximately half-way between Stoke-on-Trent railway station and the Harecastle canal tunnels. They lie on the northern side of the canal and are still named. A fine walk can be taken from Stoke-on-Trent railway station up the canal and over the Harecastle tunnels to Kidsgrove, past these warehouses and pottery works. In Ray Quinlan's book *Walking Canals in the Midlands*, there is a helpful guide.

Porthill House/Gardens. OS Sheet 118 Porthill House in Norton-in-the-Moors became the home of Hugh Henshall's nephew, Hugh Henshall Williamson. He married Anne Clowes, the great-niece of Josiah Clowes, and eventually became High Sheriff of Staffordshire in 1834. The house has been demolished but it is still possible to walk through the grounds and enjoy what is left of their ornamental gardens, **Ref 857488**.

Castlefield Warehouse, Manchester. OS Sheet 109 This warehouse in Manchester, named Henshall's Warehouse and later Grocer's Warehouse, was demolished after the Second World War. Hugh Henshall shared this warehouse with John Gilbert when running his trading company for the Trent and Mersey Canal. The warehouse possessed a lifting machine powered by a water wheel, devised by James Brindley, for landing coal at the end of the Bridgewater Canal. The site remains and part of Brindley's cast-iron wheel for operating the lift are preserved. There are also interesting copies of lithographs on the site plus plans depicting the warehouse and its machinery in its operating days. An excellent walk can be made around the basin, plus a visit to the Manchester Museum of Science and Industry.

Chesterfield Canal. OS Sheet 120 Locks at Turner Wood, built by Hugh Henshall after Brindley's death, **Ref 551812**. A fine opportunity exists to walk the eastern segment of the Chesterfield Canal, which has now been restored to the summit from Worksop. A modest but well-restored Henshall aqueduct is to be found at the eastern end of the canal near Worksop. The eastern portal of Norwood Tunnel, known originally as Hartshill Tunnel, is to be found at the summit of the canal, **Ref 500826**. An instructive walk is to be had from Worksop to Kiveton taking in the restored section with its staircase locks, restored bridges and aqueducts. Walkers can then take a train back from Kiveton station to Worksop. The western portal of Norton Tunnel is less accessible but it can be reached by walking the canal towpath from the Chesterfield Canal up to the summit.

Bolton and Bury Canal, Manchester. OS Sheet 109 Hugh Henshall surveyed and engineered this canal, starting work in 1790. The line linked the River Irwell in Manchester with the growing cotton town of Bolton. He was assisted in this work by Matthew Fletcher, a local colliery owner. To visit this abandoned waterway, an interesting section is recommended at Prestolee, **Ref 752063**, where the canal is still watered. There is a canal over-bridge and a decorative aqueduct of four arches in good condition over the River Irwell. Milestones recording the distance from Manchester were discovered by an enthusiastic walker fifteen years ago, as well as the junction with the Bury Branch Canal and a flight of locks with further over-bridges. The milestones may have been moved, officially or otherwise. A start has been made in Manchester to restore part of this little-known waterway.

Ashford Tunnel, Brecon and Abergavenny Canal. OS Sheet 161 Hugh Henshall assisted Thomas Dadford Jnr with a survey of the Llanmarch Tramway, connecting coal and iron workings to the canal. Henshall also assisted in the survey of the canal itself and made recommendations about the length and siting of Ashford Tunnel. This tunnel, found in a beautiful tree-lined setting, is situated near the Talybont-on-Usk Aqueduct and there is a scenic, 7-mile walk that can be made from here to Brecon along the canal.

Grand Western Canal. OS Sheets 164/177 Henshall worked at William Jessop's request on the surveys of this canal which was planned to link the English and Bristol channels, thereby obviating the dangerous voyage around Land's End. Henshall's task was to evaluate whether Robert Whitworth or John Longbotham's survey was the better and report back to Jessop. Little is left of the canal which only ever reached Tiverton. The method of ascent/descent on the canal was by boat lifts, designed by James Green. Remnants of these can be seen Nynehead and Greenham as well as an aqueduct over the River Tone. Approximately 20 miles of the canal has been reopened between Taunton and Tiverton and provides a pleasant walk.

Map of manor and parish of Norton-in-the-Moors, Potteries Museum and Art Gallery, Hanley. OS Sheet 118 An unusual artefact, but well worth viewing at the Potteries Museum in Hanley, Stoke-on-Trent. This enormous map, at least 6sq. ft, was drawn after extensive surveys by Hugh Henshall of the manor and parish of Norton-in-the-Moors, belonging to Charles Bowyer Adderley. Grounds owned by Josiah Clowes are included on the map as well as the signature of Hugh Henshall with a detailed crest at the top.

A.W. Skempton and Esther Clark Wright, 'Membership of Smeatonian Society'. This short transaction of the *Newcomen Society, Vol. XLIV, 1971–1972*, is well worth reading as it records not only the information concerning Hugh Henshall but also Robert Whitworth and Thomas Dadford. The transaction gives a brief, informative background to the rise of this early Society of Civil Engineers under Smeaton, which later evolved into the Institution of Civil Engineers.

Samuel Simcock

Works 1759–1768

1759–62	First Bridgewater Canal	Carpenter, Contractor
1762	Second Bridgewater Canal	Carpenter, Contractor
1760	Trent and Mersey Canal	Engineer, Contractor
1766	Staffordshire and Worcestershire Canal	Surveyor, Engineer
1767	Birmingham Canal	Surveyor, Chief Engineer
1768	Oxford Canal	Surveyor, Chief Engineer

Life *c. 1727–1804*

St Peter's church, Prestbury. OS Sheet 118 This church lies in the middle of Prestbury, an attractive Cheshire town. The church of St Peter's has Saxon origins. Indeed, there is still a small building for worship, built by the Saxons, in the graveyard. It must have made a splendid setting for the wedding of Samuel and Esther Simcock *née* Brindley, **Ref 901770.**

Ashgrove Farm, Ardley. OS Sheet 164 From Junction 10 on the M40 take the B430 towards Middleton Stoney. Once clear of the village of Ardley, follow the road for nearly a mile until a road branching to the right leads to Ashgrove Farm, which is marked. The farm is private but the cottages at the end of the lane can be seen. A fine view of the back of the farm and its acreage can be obtained by turning next right off the B430 to Upper Heyford and the Oxford Canal.

Staffordshire Canal. OS Sheets 138/127/139 Referred to in the text as the Staffordshire and Worcestershire Canal. The canal was the western arm of the Great Cross, linking the Trent and Mersey Canal to the River Severn at Stourport. Many of Brindley's school of engineers were involved. Simcock and Dadford were in charge of laying out the canal, following Henshall's survey. The canal was 46 miles long and contained forty-three locks. The locks were standardised to the gauge of the Trent and Mersey Canal and were 74ft 9in long and 7ft wide. The first lock to be built on this navigation can still be viewed at Compton, as well as an early unlined tunnel at Kinver. At St Peter's church, Kinver, John Brindley, the potter and younger brother of James, is buried. Many attractive walks are to be made down the canal and it can be cycled by the reasonably energetic in a day.

Birmingham Canal. OS Sheet 139 To walk the only remaining part of the original Birmingham Canal, as built by Simcock and Brindley, start from Aldersley Junction, **Ref 902012**, where the Birmingham Canal meets the Staffordshire and Worcestershire Canal. An energetic walk is to be had up the twenty-one locks to the centre of Wolverhampton. At the top lock are two canal houses and the Broad Street Basin. Brindley and Simcock's old canal can be followed to Horsley Junction. From there until Birmingham is reached, the canal, with its broad sweep of 'cut and fill', is the work of Thomas Telford. At Smethwick, remains of Brindley's original line and Smeaton's 'Three Lock Line' can be seen at the higher levels. At Birmingham, Gas Street Basin and the basin at Brindley Wharf, once part of Newhall Basin, remain.

Oxford Canal. OS Sheets 164/151 The Oxford Canal can be joined anywhere on its course for a walk along an early contour canal. The nearest point to Simcock's home is Lower Heyford, a good place for investigating the wharf, dry dock and canal houses. Cropredy is one of many atmospheric Oxfordshire canal villages, but one of the best walks available, which demonstrates the contour canal at its most sinewy and attractive, is the summit level from Napton Locks to Claydon, **Ref 464611**. The distance of the summit as the crow flies is 5 miles but, as the canal meanders, it is very nearly 11 miles. This stretch of canal captures entirely the original Brindley contour canal principle, engineered by his faithful brother-in-law, Samuel Simcock. Profits must have travelled slowly here.

St Mary's church, Ardley. OS Sheet 164 The burial place of both Samuel Simcock and his wife Esther is in the churchyard of St Mary's church, Ardley, **Ref 542273**. This picturesque church with Saxon origins lies on the right of the village when travelling towards Ashgrove Farm from Junction 10 on the M40; on the B430 towards Middleton Stoney. Sadly, the oolitic limestone on the graves has weathered badly and the writing on many of the older tombstones is indecipherable. At the time of writing, although the place of burial of the Simcocks at St Mary's can be verified, the actual tomb has not been identified. However, the graveyard and church are worth visiting as is the Fox and Hounds public house opposite.

Robert Whitworth

Works 1767–99

1767–69	Staffordshire and Worcestershire Canal, Birmingham Canal, Trent and Mersey Canal (part), Coventry Canal (part) and Droitwich Canal	Brindley's Chief Surveyor and Draughtsman
1774–84	Thames Navigation, improvements from Mortlake to Staines	Chief Engineer
1785–91	Forth and Clyde Canal, extension from Stockingfield–Bowling on the River Clyde, and Hamiltonhill–Port Dundas	Chief Engineer
1790–99	Leeds and Liverpool Canal, between Holmebridge near Gargrave and Henfield near Accrington	Chief Engineer
1794–97	Ashby Canal	Chief Engineer (dismissed)
1795–99	Dearne and Dove Canal	Chief Engineer
1795–99	Herefordshire and Gloucestershire Canal	Chief Engineer

NB: Forty plans and surveys were published throughout his career.

Life 1734–99

Sowerby Chapel. OS Sheet 104 The site of Robert Whitworth's christening, marriage and burial is likely to have been in the centre of Old Sowerby. The chapel, now demolished, has a graveyard which can still be seen, although it is padlocked, **Ref 036238**. No trace of Robert Whitworth's grave remains but there are Whitworths buried there from the nineteenth century who are possibly descendants. To get to the cemetery, turn left on Sowerby New Road from the A58 towards Sowerby. This road becomes Town Gate Road which in turn becomes Dob Lane. At the junction of Dob Lane with Well Head Lane, adjacent to the Rushcart Inn, the old cemetery is to be found. The village is worthy of a visit with magnificent views of Calderdale. It gives an idea of the terrain in which young Robert grew up and which must have informed his understanding of landscape. It is also on the way to Wheatley Royd House.

Wheatley Royd House, Sowerby. OS Sheet 104 Wheatley Royd House is the likely birthplace of Robert Whitworth. It was certainly where he grew up. It can be found by taking a right turn off Scout Road, having left Old Sowerby to the west. After approximately 1 mile turn right and descend the steep sides of the valley down Blind Lane. Halfway down Blind Lane, turn right and descend further along a very narrow cobbled road towards the valley floor and the embankment carrying the Manchester–Leeds Railway. Wheatley Royd House lies at the foot of the valley attached to a larger group of cottages, **Ref 026255**. These buildings were possibly the workplace of Henry Whitworth, Robert's father.

Gargrave Aqueduct, Leeds and Liverpool Canal. OS Sheet 103 The entire summit of the Leeds and Liverpool Canal is worth walking and there is a good guide in Ray Quinlan's book *Canal Walks of England and Wales*, published in 1992. To the east of the summit, after leaving Gargrave, is a very fine three-arch aqueduct over Eshton Brook, **Ref 920540**. To the west of this aqueduct a flight of locks takes the canal to its summit level before it descends to Barnoldswick, via Newton Bank Locks and the double-arch bridge at East Marton. At Greenberfield Lock the original Whitworth line with its staircase locks, now buried, has been re-routed and both lines can be seen.

Foulridge Tunnel and warehouses. OS Sheet 103 The eastern end of the tunnel can be found in the village of Foulridge, **Ref 890425**. Apart from 600yds, Whitworth constructed the tunnel using cut and cover methods. The canal company asked Josiah Clowes to check Whitworth's work which he did, saying it was cheaper and more efficient to complete the present works than to re-engineer parts of the length of tunnel. At the eastern end is a fine set of canal buildings, including Canal House, and a newly refurbished warehouse that now provides meals for visitors.

Burnley Embankment. OS Sheet 103 This magnificent piece of engineering, showing Robert Whitworth at the height of his powers, is best viewed from the bridge at the southern end of the embankment, **Ref 840325**. The bridge carries the road that leads from Burnley to Worsthorne. A dramatic walk can be enjoyed along the embankment, the 'Straight Mile', above Burnley. The town's famous football club, at Turf Moor, can be viewed from the structure, which now includes a modern aqueduct. The embankment demonstrates that 'cut and fill' techniques were certainly not beyond members of the Brindley school.

Kelvin Aqueduct, Glasgow. OS Sheet 64 This magnificent three-arch aqueduct is 400ft long and 70ft high, and is to be found at the foot of the Maryhill Locks, accessible off the A81 in Glasgow, **Ref 562690**. When opened, it was the longest aqueduct in Britain. At the foot of the locks, park and turn right. Walk for a few hundred yards below the canal, but parallel, before forking to the left under the aqueduct. Trees and foliage obscure both sides of the aqueduct, particularly the southern side, but the grandeur of the structure with its buttressing, similar to the original Barton Aqueduct, can be appreciated. Walk back to the locks and then walk on to the towpath and over the aqueduct to get some impression of its height and scale.

Maryhill Locks, Glasgow. OS Sheet 64 This flight of broad locks takes the canal from Kelvin Aqueduct up to the level that carries the waterway through Glasgow. The double locks have side pounds to provide extra water and are in good condition following refurbishment of the canal in 2001, **Ref 564690**. A short walk along the canal towards Glasgow takes you to the single-arch aqueduct over the A81.

Port Dundas Basin, Glasgow. OS Sheet 64 After the aqueduct over the A81, walk approximately 300yds down the canal and take the right arm at the junction. This arm wanders to a broad basin north of Partick Thistle Football Stadium. Continue up the arm to the basin at Port Dundas. The basin, and the locks that lead to it, are presently under refurbishment with grants of European money. A handsome Georgian Canal Company House is to be found one lock below the basin, **Ref 589665**. On a fine day, the basin affords marvellous views of Glasgow.

Bowling Locks, Junction with the Clyde. OS Sheet 64 At Bowling in Dunbartonshire the Forth and Clyde Canal joins the River Clyde. Originally there were two locks, only one of which is operational since the canal's restoration in 2001. The basin, wharves and entrance to the canal are an interesting place to visit with various eighteenth-century canal buildings and nineteenth-century bridges, **Ref 450736**. The locks into the Clyde could well be eligible for the locks with the longest balance beams in the country. A good view of the new road-suspension bridge over the Clyde is also available. Sadly, there is no memorial to Robert Whitworth, Robert Mackell or John Smeaton, and no information regarding the history of this great canal.

A mile to the east of the Bowling entrance lies Old Kilpatrick church where Robert Whitworth Jnr married Jane Fleming in 1793. Living here, Robert Whitworth Jnr was in an ideal position to contribute to the building of the western section of the canal to Glasgow, including Port Dundas.

The Wiltshire and Berkshire Canal. OS Sheet 173 Little is left of this long, winding rural canal but there are plans for restoration with individual groups working on various sections, including the branch of the North Wiltshire Canal to Cricklade where there was a junction with the Thames and Severn Canal. Near Wootton Bassett is Vastern Wharfe, **Ref 054817**, which can be visited. Here there is a wharfinger's house, cottages and a bridge over the canal. At Dauntsey Lock, **Ref 995803**, there is canal-side hamlet with the Peterborough Inn, another wharfinger's house, a lock and a stretch of canal, some of it watered. At Foxham the lock is under reconstruction and there is a new lifting bridge on a watered section that has been restored. Foxham is to be found from the B4069. Once past Foxham Inn, turn left for a few yards at the T-junction before picking up the sign on the left to the lock.

Josiah Clowes

Works 1777–94

1777	Trent and Mersey Canal	Contractor, other work not known before
1778	Chester Canal	Engineer, repairs to locks at Beeston, dismissed
1778	Stroudwater Navigation	Engineer, contract for lock repairs
1783–89	Thames and Severn Canal	Chief Engineer
1789–90	River Thames	Engineer, lock repairs
1789–92	Dudley Canal and Tunnel	Chief Engineer
1791–94	Herefordshire and Gloucestershire Canal	Chief Engineer
1791–94	Worcester and Birmingham Canal	Consultant Engineer
1792–94	Shrewsbury Canal	Chief Engineer
1792–94	Dudley No.2 Canal	Chief Engineer, Consultant
1792–94	Stratford Canal	Chief Engineer

Life 1735–94

St Bartholomew's church, Norton-in-the-Moors. OS Sheet 118 Follow the A53 north-east from Cobridge towards Endon. Approximately 2 miles from Cobridge, take the road left to Norton and Whitfield. This ascends a steep scarp. Norton church is on the left after three quarters of a mile, **Ref 894515**. Here Josiah Clowes, aged sixteen, was baptised into the Anglican faith with his two older brothers, William and John. It was also here, in 1762, Clowes married his first wife, Elizabeth Bagnall. After barely eight weeks he returned to bury her in the graveyard. Clowes and his second wife, Margaret, were buried with Elizabeth after their deaths in 1794 and 1795 respectively. Their joint tomb is to be found behind the church a few yards to the south east. The inscriptions are not wearing well and there is a need to preserve the tomb as a monument.

To the left of Josiah Clowes and his wives is the grave of his brother William, who died in 1782, and his wife Jane, *née* Henshall. Sadly, this tomb has been broken and also requires attention. Superb views are afforded from the front of the church across the Churnet Valley and Caldon Canal.

Trent and Mersey Canal, Middlewich Interchange. OS Sheet 118 Top lock, Middlewich, **Ref 668702**. The Roman town of Middlewich grew after the building of the Trent and Mersey Canal. It became important as a transhipment point from narrow to broad barges. Later, the Shropshire Union Canal joined the Trent and Mersey south of the town. Josiah Clowes was resident in Middlewich for some time. In 1781 he was living in Dog Lane, and Middlewich is recorded as his dwelling place at the time of his death fourteen years later. Clowes may have moved here whilst he was trading on the Trent and Mersey Canal, possibly for Hugh Henshall, but also in his own right. He loaned barges to the Chester Canal in 1780. It was whilst Clowes was living in Middlewich, presumably with his wife Margaret, that he returned to canal engineering, firstly with the Chester Canal and subsequently with the Thames and Severn Canal.

Stroudwater Navigation, Company House, junction with Thames and Severn Canal. OS Sheet 162 The company house of the Stroudwater Navigation is to be found near the junction of the Thames and Severn Canal and the Stroudwater Navigation at Wallbridge Locks. In Stroud, take the A46 south of the station towards Nailsworth, **Ref 850051**. Following an awkward junction where the A419 crosses over the canal, the offices are to be found on the right, to the east of Wallbridge Locks. It is probably better to park in Stroud and walk. The company offices have a fine Georgian façade but the building behind is surprisingly modest. These canal buildings were designed by local builders and this port building is very reminiscent of the Thames and Severn properties that existed at Cirencester and Cricklade. A series of walks can be taken from here west along the Stroudwater Navigation, now under restoration, or east towards Brimscombe Port on the Thames and Severn, the interchange basin from broad trows to narrowboats. The basin has been filled in but on the wall of one of the buildings is a plaque recording the site of Brimscombe Port, headquarters of the Thames and Severn Canal Company.

Round Lock Keepers House, Chalford, and walk to Red Lion Lock. OS Sheet 163 From Stroud take the A419 Cirencester road east towards Brimscombe and Chalford. Before the road crosses the canal, which is culverted at this point, there is one of the canal's five round houses, which is now a private dwelling. These tower houses were built for watchmen or lockkeepers and had three floors. From the top down there was a bedroom, on the second floor a living room and a stable at the base.

Many mills line the valley throughout. At Chalford, walk east on the towpath. At Red Lion Lock, Clowes' name is inscribed on the keystone of the lock bridge. Further up the canal, now barely watered, lies Bakers Mill Lock where there is a fine mill house and pond.

Bakers Mill Lock to west portal of Sapperton Tunnel. OS Sheet 163 From Bakers Mill Lock, ten more locks carry the canal through Frampton Mansell to the summit at Sapperton, 241ft above sea level. Some of the locks have been adapted with the unusual feature of an extra pair of gates to shorten them. This alteration in the nineteenth century was for smaller vessels. It was a desperate attempt to save water. At Daneway, the remains of a basin are to be found and the tunnellers' lodgings, built in 1784. It is now an atmospheric public house. A few hundred yards along the towpath to the east there used to be a tunnel keeper's cottage and a refuge for leggers, but this has been demolished. By the remains of the cottage the western end of the great tunnel can be found with its gothic style portal, recently rebuilt. Above the steep valley are the beautiful village of Sapperton and St Kenelm's church, where the miners that built the tunnel were buried. **Ref 947033**.

Sapperton Tunnel, east Portal and Tunnel House, Coates. OS Sheet 163 The horse path from Daneway to Coates over the tunnel remains. It affords splendid views of the spoil heaps from the shafts of the tunnel. These are now made secure with beech trees and provide an interesting feature of tunnel engineering. If driving to the eastern portal, take the Coates–Tarlton road from the village of Coates and, before the road crosses the canal, turn right parallel to the canal on an unmade road. At the top of the road is Tunnel House in Hailey Wood, used by miners during the tunnel's construction. The building used to have three floors with dormitories. Unfortunately, fire destroyed the top floor in the 1950s. It is now a public house of character, **Ref 978007**.

The canal enters the tunnel to the right of the house. The eastern portal is of classical design and possesses two niches where it was planned, but never realised, to have the statues of Sabrina and Father Thames. This portal has recently been rebuilt and trips into the tunnel can be made by arrangement with the Cotswold Canal Trust; these take place once a month. To the east of the tunnel is a straight stretch, concreted at the beginning of the twentieth century by Gloucestershire County Council in an attempt to stop leakage. This is known as the King's Reach, following the visit to the tunnel by King George III in 1789.

Dudley Tunnel/Black Country Museum. OS Sheet 139 The northern portal of Dudley Tunnel is to be found at the entrance to the Black Country Museum off the A457 at Tipton, **Ref 948915**. The southern portal is to be found at the other side of Dudley, **Ref 923893**. Trips into the tunnel through Lord Ward's Canal to view the limestone caverns and Wrens Nest Tunnel, built in the nineteenth century, as well as trips of various lengths into Dudley Tunnel itself, can be arranged through the museum. Interested parties can take a turn at legging the boat through the tunnel, and on special occasions concerts are held in the underground caverns. Visitors can appreciate the combined work of the engineers Dadford and Clowes on this narrow gauge, subterranean enterprise, now the second longest workable canal tunnel, after Standedge, on the British system. The Black Country Museum is also well worth a visit.

Brandwood Tunnel, Stratford Canal. OS Sheet 139 Many of the canals built by Josiah Clowes can be found to the south of Birmingham. These include the Worcester and Birmingham Canal, the Stratford Canal and the Dudley No.2 Canal. Two tunnels that illustrate Clowes' work can be found close together. The first, short by Clowes' standards, is Brandwood Tunnel on the Stratford Canal. The eastern portal of Brandwood can be reached by walking approximately 1 mile from Kings Norton Junction, with its unusual guillotine gate, designed to regulate water usage. The western portal, **Ref 067795**, is best reached from the Brandwood estate and has a portrait of Shakespeare on its architrave.

Wast (West) Hill Tunnel, Worcester and Birmingham Canal. OS Sheet 139 The second tunnel at Wast Hill, or West Hill as it is now known, is on the Worcester and Birmingham

Canal. This is a far longer tunnel, at 2,726yds. The northern portal is at **Ref 048780**, reached after a steep cutting, excavated by John Carne's cutting machine. The southern end is north of Newhall Farm, **Ref 036758**. Strangely, the Worcester and Birmingham Canal had broad tunnels, yet narrow locks.

Gosty Hill Tunnel, Dudley No.2 Canal. OS Sheet 139 The Dudley No.2 Canal spent a great deal of its route underground and in extremely small bore tunnels. L.T.C. Rolt described the company as 'indefatigable moles'. Gosty Hill Tunnel was 557yds and the southern portal can be found at **Ref 972852**, in what was the old Combeswood works. This portal is original, the northern portal having been rebuilt by the Great Western Railway when its Stourbridge–Dudley line crossed the canal. In addition, there is a pepperpot ventilation shaft over the line of the tunnel which can still be seen. In what used to be the old Stewarts and Lloyds works, a new canal basin and works has been created at Hawne and several sections of the canal can be walked. The southern portal of the Gosty Hill Tunnel gives some idea of what the original tunnel portals built by Clowes at Lapal must have looked like.

Site of Lapal Tunnel, Dudley No.2 Canal. OS Sheet 139 Both portals of Lapal Tunnel were lost during the building of the M5 motorway, but there are ambitious plans to return the Dudley No.2 Canal to traffic. The Lapal Trust decided that the restoration of the tunnel was unfeasible and now propose a route over the top of the tunnel through Woodgate Valley Park, as well as extensive works to return Clowes' canal to Selly Oak. In the Lapal Trust's project summary, they envisage re-exposing one of the tunnel portals and having a 'short, dry walk in length', so that visitors can obtain some idea of the tunnel.

Various photographs before and after closure of Lapal Tunnel are in existence, including pictures of the tunnel with its stop lock and steam pumping engine. The latter created a flow of water through the stygian depths to aid boats in what must have been a long and claustrophobic voyage. Dudley Library possess pictures of this tunnel for those interested in studying what became one of Clowes' longest canal tunnels, at 3,795yds. The chairman and director of the Lapal Trust can be contacted via email at chairman@lapal.org.

Berwick Tunnel, Shrewsbury Canal. OS Sheet 126 The Shrewsbury Canal was originally surveyed by Clowes as a 17-mile tub-boat canal and included eleven locks, a tunnel, an inclined plane and an aqueduct at Longdon-on-Tern. Little of the canal is left, but traces can be found in the countryside east of Shrewsbury. Berwick Tunnel is 970yds in length and its northern portal is to be found south of Preston, on the Uffington road to Atcham. The tunnel was the first in the country designed with a cantilevered wooden towpath fixed to one of the retaining walls. This was dismantled in 1819. The south portal, **Ref 541112**, is to be found at the west side of the road between Uffington and Atcham. Join the towpath after the road crosses the culverted canal and, after a walk of 400yds to the north, is the handsome tunnel mouth and a lengthsman/leggers' hut. The Berwick Wharf, earmarked for development, is close by to the east.

Hadley Park Locks. OS Sheet 127 These are the only surviving canal locks with guillotine gates; built in 1793, they underline Clowes' enthusiasm for technical invention. Conventional lock gates are built at the top of the lock but the lower gate operated like a guillotine. This gate is raised vertically by chains with counterbalance weights over the tail of the lock. The guillotine lock is to be viewed by walking up the canal from the A442 road, **Ref 672133**.

Longdon-on-Tern Aqueduct, Shrewsbury Canal. OS Sheet 127 Just over 2 miles from Rodington, after crossing a three-arched aqueduct over the River Roden, the canal crosses the River Tern on the Longdon-on-Tern Aqueduct, **Ref 617157**. The original central masonry arches, built by Clowes, have been replaced by a utilitarian iron trough designed by Thomas Telford. However, Clowes' old abutments are still in evidence. The significance of this new iron aqueduct was that it signalled a new age in canal construction led by Telford, Jessop and Outram. These radical developments in the use of new material, in the shape of wrought and cast iron, led in turn to the civil engineering of the railways. The Shrewsbury and Newport Canal Trust has been set with a view to the restoration of as much of the canal as is possible.

Oxenhall Tunnel, Herefordshire and Gloucestershire Canal. OS Sheet 163 The southern remains of the Herefordshire and Gloucestershire Canal, surveyed and built by several members of the Brindley school, can be traced, in part, from the basin on the River Severn to Hereford itself. Henshall re-surveyed the canal in 1792 recommending a shorter, but more demanding, route than that originally surveyed by Josiah Clowes; the Newent Branch being incorporated into the main line. The aqueduct at Ellbrook and the locks south of Oxenhall are an interesting place to visit in order to observe the restoration that has taken place so far on the canal. The newly restored basin at Over is also worthy of attention. A walk north, up the canal from the Oxenhall lock, leads to the southern portal of Oxenhall Tunnel, **Ref 162149**.

Thomas Dadford

Works 1769–97

1769–74	Staffordshire and Worcestershire Canal	Carpenter, Contractor, Resident Engineer
1776–81	Stourbridge Canal	Chief Engineer
1776–83	Dudley Canal	Chief Engineer
1780–87	Trent and Mersey Canal	Resident Engineer
1784–87	Coventry Canal	Advice on aqueduct over River Thames and Fazeley–Fradley section of canal
1785	River Severn, Maisemore Bridge	Engineer
1789–91	Cromford Canal	Contractor, withdrew in financial difficulties
1790–94	Glamorganshire Canal	Engineer and Contractor, dismissed
1796–97	Montgomeryshire Canal	Engineer, Carrehofa to Garthmyl
1796–97	Stafford Gaol and Sedgley Park School	Architect

Life *c. 1730–1809*

St Peter's church, Wolverhampton. OS Sheet 139 The majestic church of St Peter's in Wolverhampton, **Ref 917988**, is well worth a visit. It is where Thomas Dadford and Frances Brown were married in 1759, and where they were both buried at the beginning of the nineteenth century. Sadly, developments in Queen Square have meant that their tombs have been lost. However, once the concrete collar surrounding the centre of Wolverhampton has been breached, the town with its various attractive Georgian and Victorian buildings is a pleasant surprise, in what has become a densely packed industrial and commercial area. To the north-east of the town one of last original parts of the Birmingham Canal, before Telford rebuilt it, can be viewed.

Compton Lock. OS Sheet 139 The lock, stated by Joseph Priestley to have been Brindley's first lock, is to be found approximately 2 miles south of Aldesley Junction, **Ref 885990**. The junction is where the Staffordshire and Worcestershire Canal meets the Birmingham Canal and is 2 miles due west from Wolverhampton on the A454. Once the A454 crosses the canal, walk north for approximately 400yds to the lock. This is a good section to continue walking. Beyond Aldesley Junction, ¾ mile to the north, lies Autherley Junction; the connection with the Birmingham and Liverpool Junction Canal, built by Thomas Telford.

Bratch Locks. OS Sheet 139 These attractive triple staircase locks are situated near Wombourne, **Ref 867939**. Travelling north towards Wolverhampton on the A449, turn left down the wonderfully named Billy Buns Lane which then becomes Bratch Lane. The village of Wombourne is to the south. Here the staircase locks, built by Brindley and Dadford, can be found – although today they are laid out as separate chambers. There is an interesting octagonal toll house and water-pumping station that was installed during the Victorian era, to pump much-needed water to the summit level.

Stourport Locks/Basin. OS Sheet 138 Stourport-on-Severn is worth a visit in order to see a town created by a canal. A good walk can be taken up the canal from the barge locks that connect the Staffordshire and Worcestershire Canal to the River Severn. From these locks, narrow locks lift the canal to the Lower Basin and then further staircase locks to the Clock Basin. The Tontine Hotel, near the entrance to the River Severn, was built in 1788 and was where the Staffordshire and Worcestershire Canal management committee met their shareholders.

Stourport, as a town created by the canal, was a precursor to Swindon and Crewe in the railway age and delighted Elizabeth Prowse on her pleasure-boat trip in 1774. The town can be reached from Kidderminster by taking the A451.

Sow Aqueduct (or the Milford Aqueduct). OS Sheet 127 Take the A513 from Rugeley north past Shugborough Hall and turn right to Tixhall. A fine view of the side of the aqueduct is to be obtained from the Milford–Tixhall road. A short walk from the canal bridge north of Milford takes you to the substantial four-arch aqueduct, with 22ft spans over the River Sow on the Staffordshire and Worcestershire Canal, **Ref 973215**. The aqueduct has an unusually broad towpath, characteristic of early canal aqueducts of the Brindley school. Clearly, many different designs were being tested in this early stage of aqueduct development. There is also a fine canal-side house, wharf and lock at Tixhall, to be found approximately 400yds from the road bridge to the east. The brief canal walk affords views of the Shugborough estate and the castellated tunnel portal on the Trent Valley line of the LNWR Railway.

St Cassian's church, Chaddesley Corbett. OS Sheet 139 This elegant church, dating from the twelfth century, was where Thomas Dadford was consulted about the rebuilding of the tower and the addition of a spire. John Roper, in his history of the church, thinks Dadford was probably consulted by Sir Robert Throckmorton; Mr Dadford being well known in Roman Catholic circles in the West Midlands. It was also where Thomas Dadford Jnr married Ann Parker of Bluntington Green, Chaddesley Corbett, on 15 August 1797.

The village of Chaddesley Corbett lies halfway between Bromsgrove and Kidderminster on the A448. The church is immediately on the left after the turn from the A448 to the village, **Ref 891738**. Once through the village, a right turn leads to the village of Bluntington. Harvington Hall, home of Roman Catholics during the Reformation, is close by and worth a visit.

Navigation House, Abercynon. OS Sheet 170 The headquarters of the Glamorganshire Canal, built by Thomas Dadford Snr, can be found by taking the A4054 into Abercynon from the A472. Navigation House, now not in use, is to the right before the fire station. The latter has been built on the basin of the canal where Dadford built a maintenance yard. The River Taff can be seen from the road bridge. This is where the Aberdare Canal and the Pen-y-daren tramroad joined the Glamorganshire Canal, and where the canal changed from one side of the valley to the other on an aqueduct built by Dadford. Sadly, this is long gone.

Chapel Row, Merthyr Tydfil. OS Sheet 160 Although the Glamorganshire Canal has been almost completely obliterated by the building of the A470, a monument to its existence remains near Chapel Row in Merthyr Tydfil. Part of the canal was excavated to the site of Lock 1. Today, opposite Charles Parry's birthplace in Chapel Row is a shallow ornamental basin spanned by a cast-iron canal bridge which has been resited from Rhydycar. There are also stone boundary markers from the canal with the initials G.C. These were known as tombstones, the reason being evident in the shape. There are also remnants of the flanged tramway and tramway wagons that served the canal. James Dadford, Thomas' son, hired boats and wagons to the ironmasters for work on the canal. The Dadfords built boats that Rowson and Wright comment were very similar in style to the 'starvationers', found on the original Bridgewater Canal at Worsley.

St Teilo's church, Llanarth, Thomas Dadford Jnr's grave. OS Sheet 161 South of Abergavenny take the B4598 from the roundabout with the A40. Follow the B4598 for approximately 3 miles. Take the second left from the B4598 after it crosses the A40 at Llanarth. St Teilo's church is on the right at the top of the hill. Thomas Dadford Jnr's tomb, now sadly cracked, is 30yds to the right of the path leading to the church entrance. It is in a secluded area under a yew tree. At Abergavenny his beautiful and skilfully engineered canal, the Monmouthshire and Brecon Canal, can be walked.

Sedgley Park School. OS Sheet 139 At Wolverhampton the school where three of Dadford's sons, William, James and John, were educated still exists, although it is now a Ramada Hotel in Park Drive, Goldthorne Park. From the A449 turn into Goldthorne Hill, take the third turning right into Edman Road and after 500yds the hotel is reached.

The chapel, built by Thomas Dadford Snr, is now an Indian restaurant. Apart from engineers the air in that part of the Midlands clearly favoured actors, witness David Garrick in Lichfield as well as John Phillip Kemble, who was educated at Sedgley Park School. There is a plaque in his memory but, sadly, no record of the Dadfords.

Montgomeryshire Canal, Vyrnwy Aqueduct. OS Sheet 126 Vyrnwy Aqueduct is by the junction of the B4398 and B4393, 1½ miles south-west of Llanmynech, **Ref 254197**. The aqueduct was begun by John Dadford and completed by his father, Thomas Dadford Snr, after the collapse of one of its 39ft arches. The aqueduct comprises five arches with a smaller three-arch aqueduct over the flood plains next to it. In 1823 the whole structure was repaired by George Watson Buck and iron bands were used for strengthening. Between the aqueduct and Llanmynech are the Carreghofa Locks and beyond the aqueduct, after a right-hand bend, is a warehouse.

John Varley

Works 1759–1801

c.1759	Bridgewater Canals	Working with James Brindley
1766–c.71	Trent and Mersey Canal	Working with James Brindley
1771–77	Chesterfield Canal	Resident Engineer
1777–79	Erewash Canal	Resident Engineer
1782–92	Sleaford Navigation, Nutbrook Canal, Nottingham and Beeston Canal	Surveyor, Consultant
1793	Leicestershire and Northamptonshire Union Canal	Resident Engineer with son
1798–1800	Wolseley Bridge, Colwich, Staffs	Contractor
1800–01	Huddersfield Narrow Canal	Subcontractor, repairs

Life 1740–1809

Much of John Varley's work can be visited when looking at the work of other members of the Brindley school. The Chesterfield Canal, where he worked under James Brindley and then Hugh Henshall, has been described. Wolseley Bridge can be viewed when visiting the Wolseley Arms, the meeting place for the promoters of the Grand Trunk Canal. Varley's son, also named John, worked with his father on the Leicestershire and Northamptonshire Union Canal, now the Leicester arm of the Grand Union Canal.

At the time of writing, John Varley's grave at Harthill has not been discovered.

Samuel Weston

Works 1759–95

c.1759	Bridgewater Canal	Staff Holder, Surveyor
1770–74	Chester Canal	Resident Engineer
	Leeds and Liverpool Canal	Contractor
1785–90	Oxford Canal	Contractor and Surveyor
1794–95	Ellesmere Canal	Contractor

Life c.1730s–1804

Again, these canals have been referred to previously under other engineers from Brindley's school. Weston's contributions can be only be guessed at when viewing these works. However, he was highly regarded by Smeaton and Jessop. His 4-mile section of the Leeds and Liverpool Canal between Newburgh and Liverpool can be walked, as well as his infamous aqueduct at Gowy on the Chester Canal. Mike Chrimes' life of Samuel Weston in the ICE Dictionary is recommended for those interested in Weston and his son William. At the time of writing, it is not known where he was brought up near Oxford, where he died or whether his grave still exists.

James Brindley Jnr

Works 1775–1800

1775–76	Ballendine's Canal, Maryland	Engineer
1783	Susquehanna Canal, Maryland	Engineer
1785–1802	Potomac Canal, Great Falls Bypass, Maryland	Engineer
1793–1800	Santee Canal, South Carolina	Engineer
1785	James River Canal	Engineer
1792–97	Conewago Canal, Pennsylvania	Engineer
1800	Harper's Ferry Power Canal, West Virginia	Engineer

Life 1745–1820

A good excuse to visit America. In addition to these canals, mentioned in greater detail by Robert Kapsch and Yvonne Long in the *International Journal for the History of Engineering & Technology* (Vol. 81, No.1, 2011), the obelisk to James Brindley and members of his family can be found at Old Swedes church, Wilmington. The village of his birth is at Waterfall in the Staffordshire Moorlands.

Apart from James Brindley, there are no monuments to the engineers recorded in this book. It is hoped that further details of lives, letters and portraits will be forthcoming from future researches by canal historians and enthusiasts. In addition, it would be helpful for the general public if plaques were installed on or near the engineers' works to record their considerable achievements. Without this information it is not surprising that many people have never heard of these engineers.

Finally, it would be appropriate for the foliage and tree growth that has occurred in front of some of these works to be cleared so that a good view may be had of the structure. A classic example would be the tree growth in front of Kelvin Aqueduct in Glasgow, one of Robert Whitworth's masterpieces.

Hugh Taylor, navigator extraordinaire, in front of the Kelvin Aqueduct, Forth and Clyde Canal.

References

Introduction

Brindley, James (see Chapter 3)
Rolt, L.T.C., *Navigable Waterways*, (Longmans, 1969)

Eighteenth-Century Canal Construction

Biddle, G., *The Railway Surveyors* (Ian Allan & British Railways Property Board, 1990)
Clarke, *Practical Tunnelling* (Crosby Lockwood & Co. 1877)
Cox, R., *Short History of Civil Engineering* (Circo, 1996)
Rees, A., 'Setting out Canals', *Encyclopaedia* (1800)
Simms, F.W., *Practical Tunnelling* (Extended, D.K. Clarke, Lockwood, 1896)

1 James Brindley 1716–72

Original Sources

References have been taken from Brindley's tomb at St James church, Newchapel. Also, references from the Henshall tombs and the tablet in St James church, Newchapel
Brindley, James, Boiler with Internal Firebox, Patent 730, 26 December 1758
———, Personal notebooks, called Day Books, Bk 1 1761–1762/Bk 2, October–November 1763. Archives of the Institution of Civil Engineers, ICE London
Wolstanton Parish Records 1624–1812 (1914–1920) SRPS 2 Vols

Published Works

Banks, A.G., Schofield, R.M., *Brindley at Wet Earth Colliery* (David & Charles, 1968)
Bode, H., *James Brindley, Lifelines 14*
Boucher, Cyril T.G., *James Brindley Engineer 1716–1772* (Goose & Son, Norwich, 1968)
Boyes, John H., 'James Brindley', *Biographical Dictionary of Engineering*. Vol.1. 1500–1830 (2002)
Darwin, Erasmus, 'Eulogium', *The Economy of Vegetation* (1772)
Evans, Kathleen, *James Brindley: A New Perspective* (Churnet Valley Books, 1997)
Hitchings, Henry, *A Dictionary of English Language, Dr. Johnson* (John Murray, 2005)
Johnson, Samuel, *A Dictionary of the English Language*, 2 Vols (1755, 1979 Edition, Times Books Ltd)
King Head, Desmond, *Erasmus Darwin: A life of unequalled achievement* (1999)

Kippis, 'James Brindley. Material from Mr Henshall with character sketch by Mr Bentley', *Biographica Britannica*

Klemperor, W.D., and Sillitoe, P.J., *James Brindley at Turnhurst Hall.* No.6 Staffordshire Archaeological Studies. City Museum and Library, Stoke-on-Trent (1995)

Lead, P., *Agents of Revolution – John and Thomas Gilbert – Entrepreneur.* Centre for Local History, Keele (1989)

————, *The Trent and Mersey Canal* (Moorland Press, 1980)

Lindsay, Jean, *The Trent and Mersey Canal* (David & Charles. 1979)

Malet, Hugh, *Bridgewater, The Canal Duke* (Manchester University Press, 1977)

Meteyard, Eliza, *Life of Josiah Wedgwood* (Reprint Press 2 Vols. 1970. Original published 1865)

Priestley, J., *Historical Account of the Navigations, Rivers, Canals and Railways throughout Great Britain* (1830)

Richardson, Christine, *James Brindley: Canal Pioneer* (Waterways World, 2004)

Rolt, L.T.C., *From Sea to Sea, The Canal du Midi* (Allen Lane, 1973)

Rudkin, Peter Cross, 'Constructing the Staffordshire and Worcestershire Canal, 1766–72, *Newcomen Society Transactions*, Vol.75, No.2 (2005)

Saul, A.R., 'James Brindley and his Staffordshire Associates' (1939)

Smiles, Samuel, *Lives of the Engineers*, Vol.1 (1861)

2 Hugh Henshall 1734–1816

Original Sources

References have been obtained from gravestones at St James church, Newchapel, where the Henshall family are buried next to James Brindley. Other information has been gathered from Norton-in-the-Moors churchyard where William Clowes and Jane Clowes, neé Henshall, are buried. Josiah Clowes, the canal engineer and brother-in-law of Hugh Henshall, is also buried in this churchyard.

Aris's Birmingham Gazette, 12 June 1775, Advertisement for Labourers on the Trent and Mersey Canal

Boat hired, 1735/1779–1791, Accounts and advice, 21 items. Hugh Henshall and Co. and Josiah Wedgewood. Wedgwood Papers, University of Keele

Burslem Parish Records, Parts 1–3, 1578–1812 (SPRS 1913)

Extract from indenture between Dorothy Norton and Hugh Henshall, BWA, Ref. BW No.1105.95

'Indentures/Hugh Henshall Esq. and others to Messrs Bancks and Barker/Lease for ironstone mining at Golden Hill. Cancelled by Henshall, 1820, reinstated by Anne Brindley et al, 1840' Spode Collection, University of Keele, Ref. 229g

Letter from E. Lingard to Hugh Henshall, British Waterways Archive; Ref. BW No.1080.95

Letter from Hugh Henshall to W. Harrison, 10 May 1773, BWA, Ref. BW No.1127.95

'Memo of an Agreement', 1 February 1767, between George and John Whiston, Sarah Barker and William Cross. 'Purchase of lands at Kings Bromley', BWA, Ref. BW No.1061.95

'Memo of an Agreement', 21 July 1767, between N. Kent and John Sparrow and Hugh Henshall, BWA, BW No.1058.95

Minute Books of the Chester Canal Company, PRO RAIL (3 December 1779, 20 January 1780)

Minutes of the Chesterfield Canal Company, PRO Ref. 817/8 (January/April 1774)

Mortgage of Greenway Bank, 1809, Staffordshire Record Office, Stafford
Norton-in-the-Moors, Parish records, Part 1, 1574–1751 (Staffordshire Parish Record
    Society (SPRS) 1754/1837)
Norton-in-the-Moors Parish Records, Part 2, 1754–1837 (SPRS 1942/3)
'Plan of the manor and Parish of Norton-in-the-Moors belonging to Charles Bowyer
    Adderley with his estates lying thereon, surveyed Hugh Henshall', City Library, Hanley,
    Stoke-on-Trent
Stoke-on-Trent Parish Records, Parts 1–4, 1624–1812 (SPRS 1914, 1918, 1925 and 1926–27)
'Will of Hugh Henshall of Longport, Burslem', PRO Ref. PROB 11/1593 CP 3438
Wolstanton Parish Records, Parts 1–2, 1624–1812 (SPRS 1914)

Published Works

Dolan, B., *Josiah Wedgwood, Entrepreneur to the Enlightenment* (Harper Perennial, 2004)
Evans, K.M., *James Brindley, Canal Engineer: A new Perspective* (Churnet Valley Books, 1997)
Greenslade and Jenkins, *Victoria History of Stafford*, Vol.2
Hadfield, C., *Notes on individual engineers* (LSE)
———, *Canals of the West Midlands* (David & Charles, 1969)
———, *Canals of the East Midlands* (David & Charles, 1970)
———, *Canals to Stratford* (1966)
———, *Canals of South Wales and the Border* (1967)
Hadfield, C., and Biddle, G., *The Canals of North West England*, 2 Vols (David & Charles, 1970)
Holgate, D., *Newhall and its Imitators* (Faber & Faber, 1987)
Household, H., *The Thames and Severn Canal* (Alan Sutton, 1969)
Lead, P., *The Caldon Canal and Tramroads* (Oakwood, 1990)
———, *Agents of a Revolution, J. and T. Gilbert, Entrepreneurs* (1989)
———, *The Trent and Mersey Canal* (Moorland, 1980)
Lewis, C.G., 'Josiah Clowes', *Newcomen Society Transactions*, Vol.50 (1978–79)
———, 'Hugh Henshall', *Newcomen Society Transactions*, Vol.6 (2006)
———, 'Summit Problem', *Waterways World*
Lindsay, J., *The Trent and Mersey Canal* (David & Charles, 1979)
Malet, H., *Bridgewater, The Canal Duke 1736–1803* (David & Charles, 1977)
Meteyard, Eliza, *The life of Josiah Wedgwood*, 2 Vols (1865)
Phillips, John, *General History of Inland Navigation Foreign and Domestic* (1795)
Ransom, P.J.G., *The Archaeology of the Transport Revolution, 1750–1850* (World's Work, 1984)
Skempton, A.W., and Wright, H.C., 'Early Members of the Smeaton Society', *Newcomen
    Society Transactions*, Vol.44 (1972)
Smiles, S., *Lives of the Engineers*, Vol. 1 (Murray, 1861)
Thomson, N.A., 'The renovation of Ashford Tunnel', in Schofield, R.B., *Benjamin Outram*
    (Merton Press, 2000)
Turnbull, G.L., *Traffic and Transport, An economic history of Pickfords* (1979)

3 Samuel Simcock c.1727–1804

Original Sources

Brindley, James, Personal Day Book, October–November 1763. MS 85/2/42, Archives ICE,
    London
Letter from E. Lingard to Hugh Henshall. British Waterways Archive, Ref. BW, No.1080.95
Parish Registers/Deaths church of England, Parish church of Ardley

Simcock, S., Signature Cherwell Aqueduct. Warwickshire County RO. CR1590/P2,3. 1786
Will of Samuel Simcock Jnr, 1817. Proved 19 March 1805. PRO Cat Ref. Prob 11/1631
Will of Esther Simcock. PROB 11/1483
Will of Brindley Snr M/S

Published Works

Angerstein, R.R., *Roads and Traffic, Illustrated Diary 1753–1755* (Newcomen Society)
Boucher, Cyril T.G., *James Brindley Engineer 1716–1772* (Goose & Son, Norwich, 1968)
Broadbridge, S.R., *The Birmingham Canal Navigations, Vol.1 1768–1846* (David & Charles, 1974)
Broom, C.I., 'The Western Canal, Forerunner of the Kennet and Avon Canal', *Newcomen Society Transactions*, Vol.80, No.1 (2010)
Chrimes, M., *Biographical Dictionary of Engineering*, ICE, Vol.1 (2002)
Collins, P., *Black Country Canals* (Sutton Publishing, 2001)
Compton, Hugh, *The Oxford Canal* (David & Charles, 1976)
Cross Rudkin, Peter, 'Constructing the Staffordshire and Worcestershire Canal, 1766–72', *Newcomen Society Transactions*, Vol.75, No.2 (2005)
Hadfield, Charles, *Canals of the West Midlands* (David & Charles, 1969)
———, *Canals of South and South-East England* (David & Charles, 1969)
Phillips, John, *A General History of Inland Navigation* (David & Charles, 1970)
Rees, Abraham, *Encyclopaedia, Canals* (1800)
Richardson, Christine, *James Brindley, Canal Pioneer* (Waterways World, 2004)
Victoria History of Oxfordshire, Vols 2, 6, 9, 10

4 Robert Whitworth 1734–99

Original Sources

Minute Book of the Smeatonian Society, 1771–1774 (ICE)
'Obituary', *Newcastle Advertiser* (20 April 1799). Colindale Ref. M35533-35
Parish Registers of Halifax and Sowerby, West Yorkshire Archive Service
Whitworth, Robert, 'Plan for Estimates of intended Navigation from Lough Neagh to Belfast' (1770). Ref. WH!/PE1, T8V/73. ICE
———, 'Plan of intended navigation canal from Coventry near Griff in the county of Warwickshire to the coal mines at Measham, Oakthorpe, Donisthorpe and Ashby Wolds in the counties of Leicester and Derby'. Ref W/781/WH1 PIN. ICE
———, 'Report and survey of the canal from Waltham Abbey to Moorfields' (1773). Ref. WH1 RSC, TFV/1
———, 'Plan of navigation, canal from Andover to Redbridge in the County of Southampton' (1770), *Gentleman's Magazine* (8 February 1772). Ref. G772 GEN, PPN. ICE
———, 'Plan & estimate of intended navigation from Lough Neagh to Belfast' (1770). Ref. / WH1 PE1
'Will, Robert Whitworth', PRO Proved April 1799

Published Works

Clarke, M., *Leeds and Liverpool Canal*
Dalby, L.J., *The Wilts and Berks Canal* (Oakwood Press, 1971)

Hadfield, Charles and Biddle, Gordon, *The Canals of North West England*, 2 Vols (David & Charles, 1970)

Hadfield, C., *Canals of the West Midlands* (David & Charles, 1969)

———, *Canals of the East Midlands* (David & Charles, 1970)

———, *Canals of South Wales and the Border* (David & Charles, 1967)

———, *Canals of South and South-East England* (David & Charles, 1969)

Hitchings, Henry, *A Dictionary of the English Language, Dr Johnson* (John Murray, 2005)

Household, Humphrey, *Thames and Severn Canal* (David & Charles, 1969)

Lindsay, Jean, *The Canals of Scotland* (David & Charles, 1979)

Oxley, G.W., 'Robert Whitworth', *Biographical Dictionary of Engineers*, Vol.1 (ICE, 2002)

———, 'Robert Whitworth 1737–99. Canal Engineer of Calderdale', *Transactions of the Antiquarian Society*, Vol.8 (New Series, 2000)

Trigg, W.B., 'The Canals and Waterways', *Stories of the town that Bred us*, Compiled by Mulroy, J.J. (Halifax, 1948)

5 Josiah Clowes 1735–94

Original Sources

Advertisement for labourers on the Trent and Mersey Canal, *Aris's Birmingham Gazette* (12 June 1775)

Burslem Parish Records, *SPRS*, 1578/1812, Vols 1–3 (1913)

Dudley Canal: Old Tunnel (Notes prepared by Railway and Canal Historical Society.) C. Weaver: B.T.H.R. PRO

Document concerned with the transfer of property by 'William Clowes of Porthill in the parish of Wolstanton, potter, nephew and devisee, named and appointed in the will of Josiah Clowes, late of Middlewich in the County of Chester, deceased'. The Adams Collection, City Central Library, Bethesda Street, Hanley, Stoke. Doc. Ref. EMT8/796

Hadfield, C., 'Notes on individual engineers'. Kindly lent by the late Charles Hadfield, London School of Economics

Josiah Clowes' Will, Proved 1795. Cheshire R O. Ref. WS. 1795

Land Tax Assessment, Middlewich. Cheshire R O. Ref. QDV 2/288

Letter by Smeaton, 22 July 1791, re: Worcester and Birmingham Canal

Letter from John Davenport to Thomas Clifford, 22 September 1773, about Clowes quitting the Staffordshire and Worcestershire Canal

Manuscript Evidence on Private Bills. 1791 Birmingham canal (C.W.H.) House of Lords.

Minute Books of the Chester Canal Company, B.T.H.R. PRO Entries for: 20 September 1776; 29 January 1777; 24 April 1778; 30 October 1778; 3 December 1779; 20 January 1780; boat hire

Minute Books of the Huddersfield Canal Company. Entries from 26 June 1794–24 June 1813. (No mention of Josiah Clowes)

Norton-in-the-Moors Parish Records, *Staffs. Parish Record Society* (SPRS), 1574/1751, Vol.1 (1924); 1754/1837, Vol. 2 (1942/3). (These transcripts are incomplete)

Norton-in-the-Moors Parish Records for record of deaths after 1794. (The registers are in the church. Marriages are in a separate volume.)

Rate books, Castle Ward, Cirencester, 1788/9. Shire Hall, Gloucester

Report of Clowes' death, *Gloucester Journal* (Monday 9 February 1795)

'Report of Company to General Assembly of Hereford and Gloucester', 10 November 1796, shows Clowes' report and estimate was not considered altogether satisfactory

Reports and correspondence between Josiah Clowes and the Thames and Severn Company, June–July 1789. Ref. TS 194/4, Shire Hall, Gloucester

State of Accounts of the Herefordshire and Gloucestershire Canal Company, 7 November 1796, shows Clowes held four shares, £520 fully paid up

Stoke-upon-Trent Parish Records, *SPRS*, 1629/1812, Vols 1–3 (1914/27)

Stroudwater Navigation minute book. Ref. 8 April/21 May 1794

Thames Navigation Minute books, 1 July 1789/14 July 1790 as in source 21. Berkshire County Records, Reading

William Clowes' Will, Proved 1823, with codicil. Lichfield Joint Record Office

Wolstanton Parish Records, *S.R.P.S.*, 1624/1812, 2 Vols (1914/20)

Published Works

Cragg, R., *Civil Engineering Heritage, West Midlands* (Phillimore & Co., 2010)

Duncan Young, W.E., *The Thames and Severn Canal* (Gazebo Press, 1968/69)

Field, R.D., *The Grand Scheme* (Printed by W. Langsbury, Cheltenham (n.d.))

Greenslade and Jenkins (eds), *Victoria History of the County of Stafford*, Vol. 2 (Oxford University Press, 1967)

Hadfield, C., *Canals of South Wales and the Border* (David & Charles, 1967)

———, *Canals of the West Midlands* (David & Charles, 1966)

Hadfield, C. and Biddle, G., *Canals of North West England* (David & Charles, 1970)

Hadfield, C. and Norris, *Canals to Stratford* (David & Charles, 1962)

Holgate, David, *Newhall and its Imitators* (Faber & Faber, 1971)

Household, H., *The Thames and Severn Canal* (David & Charles, 1969)

Lewis, C.G., 'Josiah Clowes, Part 1', *Waterways World*, No. 79 (April 1978)

———, 'Josiah Clowes, Part 2', *Waterways World*, No. 80 (May 1978)

———, 'Josiah Clowes', *Newcomen Society Transactions*, Vol. 50 (1978–79)

———, 'Summit Problem. Sapperton Tunnel', *Waterways World*, No. 100 (May 1980)

Priestley, J., *Historical Accounts of the Navigable Rivers, Canals and Railways throughout Great Britain* (1830)

Rolt, L.T.C., *The Life of Thomas Telford* (Longmans, 1958)

———, *Navigable Waterways* (Longmans, 1969)

Russell, R., *The Lost Canals of England* (David & Charles, 1971)

Smiles, Samuel, *Lives of the Engineers*, Vol. 1 (1861)

6 Thomas Dadford c.1730–1809

Original Sources

Tombstone, St Teilo's church, Llanarth. Thomas Dadford Jnr's Grave

Register and Marriage Bonds. St Peter's Wolverhampton. *c.*1750, Lichfield RO

Marriage Register, Belbroughton, *c.*1797. Worcestershire RO

Letter from Robert Thompson to William Tait, 23 March 1792. Glamorgan RO 1782–94

Staffs and Worcs, Order Book 1766–1785, 21 April 1774, Dwelling Thos. Dadford

Staffs and Worcs Canal, Minutes, 17 March 1767

'Will of Thos. Dadford', PRO, PROB, 11/1509

'Obituary of Thos. Wedgwood', *Staffordshire Observer*, Lichfield RO

Minutes of the Smeatonian Society, 25 April 1783 (ICE)

Hadfield, C., 'Index of Canal Engineers, Surveyors and Contractors' (LSE, 1967)

Norris, J., 'Extensive notes on Thos. Dadford and Sons'

Published Works

Cross Rudkin, Peter, 'Constructing the Staffordshire and Worcestershire Canal, 1766–72', *Newcomen Society Transactions*, Vol.75, No.2 (2005)

Hadfield, C. and Skempton, A.W. *William Jessop, Engineer* (David & Charles, 1981)

Hadfield, C., *Canals of the West Midlands* (David & Charles, 1969)

———, *Canals of the East Midlands* (David & Charles, 1970)

———, *Canals of South Wales and the Border* (David & Charles, 1967)

Hughes, S., 'The Archaeology of the Montgomeryshire Canal', *Royal Commission on Ancient & Historical Monuments in Wales* (1988)

Longford, J.I., 'Dudley Tunnel, a history', *TRAD guide* (Dudley Canal Trust, 1973)

Longford, J.I., *Stourbridge Canal/Towpath Guide* (Wolverley Press, 1992)

Norris, J., *History of Brecknock and Abergavenny Canals*, Part 1 (J. Norris, 2010)

———, 'Life of Thos. Dadford Senr./Thos. Dadford Junr./James Dadford/John Dadford', *Biographical Dictionary of Civil Engineers, 1500–1830*, Vol.1 (ICE, 2002)

———, *The Monmouthshire and Brecon Canal* (J. Norris 1991, revised edition 2010)

Roper, J.A., *History of St Cassian's Church, Chaddesley Corbett* (Friends of St Cassian's church, 2006)

Rowson, S., and Wright, I., *The Glamorganshire and Aberdare Canal*, 2 Vols (Black Dwarf, 2001)

Schofield, R.B., *Benjamin Outram, an Engineering Biography* (Merton Priory Press, 2000)

7 John Varley, Samuel Weston and James Brindley Jnr

Barton, B., 'William Weston', *Biographical Dictionary of Civil Engineers* (ICE, 2002)

Chrimes, M., 'Samuel Weston', *Biographical Dictionary of Civil Engineers* (ICE, 2002)

Emery, G., *The Old Chester Canal, History and Guide* (G. Emery, 2005)

Hadfield, C., *Canals of the West Midlands* (David & Charles, 1969)

———, *Canals of South and South-East England* (David & Charles, 1969)

———, *Canals of North West England*, 2 Vols (David & Charles, 1970)

Kapsch, R.J. and Long, Y.E., 'James Brindley, American Canal Engineer', *International Journal for the History of Engineering and Technology* (2011)

Richardson, Christine, 'John Varley', *Biographical Dictionary of Civil Engineers* (ICE, 2002)

———, *The Waterways Revolution – From the Peaks to the Trent, 1768–1778* (Hartnolds Ltd, 1992)

'The Story of James Brindley, Canal-Engineer, 1745–1820', *Inland Waterways Protection Society*, Newsletter 174 (2003)

Will of Samuel Weston, Proved 19 March. 1805PRO. Cat Ref. Prob 11/1423

Conclusion

Chrimes, M., *Civil Engineering 1839–1889, A Photographic History* (Alan Sutton in association with Thomas Telford Ltd, 1991)

———, 'Samuel Bull', *Biographical Dictionary of Civil Engineers* (ICE, 2002)

Index